THE

Velocette

SAGA

THE STORY OF A GREAT BRITISH MOTORCYCLE

C. E. - 'Titch' - ALLEN

About the author

Titch Allen (C.E. Allen, BEM; the medal was awarded for his service as a DR in 1940-46) found that his original job as a newspaper reporter got in the way of his love of motorcycle sport, so he swapped roles by entering the motorcycle industry on the accessory side. Sponsored by the entrepreneurial Jim Ferriday, founder of the Feridax firm, he carried out a number of long-distance publicity rides, including a record-breaking 3,600 miles in 10 days circuit of the coast of England, Scotland and Wales under ACU observation on an Ariel Huntmaster outfit. His business travels gave him opportunity to promote the Vintage Motor Cycle Club which he had founded in 1946 after laying its foundations during the war.

Writing now became a hobby to be enjoyed when the sprit moved him and the appearance of *Motorcycle Sport* in 1962 gave him a platform to air his views and pass on his wide knowledge of vintage and classic motorcycles. Since then he has been a regular contributor.

In 65 years of motorcycling (he began off road when a schoolboy), he has owned or ridden most makes and admits to love affairs with one or two while strenuously denying that he is a one-make enthusiast. He likes to be thought of as an all-makes enthusiast... how else, he asks, can you appreciate one machine if you have not sampled the rest? But a first love is seldom forgotten, and he has never forgotten the 1930 'cammy' model he owned in his teens. "The first really good machine I owned", he says. Tempted by more sophisticated bikes with more gears (the vintage Velo only had three) and electric lights (his Velo had gas), he deserted his first love but never forgot it. So when Vintage racing took off in the sixties and his eldest son wanted to compete, Dad built up (with the aid of Ivan Rhodes) a KSS bitza which was almost identical to that first love. Son Roger moved on to mightier metal, so Dad took over the Velo and raced it for 20 years. Still rides it in sprints. Pre-war MAC and MSS models and a MkIV KTT have been raced over the years and a couple of GTP two-strokes and a Viper have been in the stable. Jewel in the crown, of course, is the Model O, the one-off Irving-designed vertical twin. A love at first ride affair, this one joins the cammy racer as his last loves.

In the face of such marque loyalty he cannot deny that he has been a Velocette enthusiast for a lifetime, still is, and always will be.

Book Editor: C.J. Ayton

Cover design: Ruth Sutherland

© Amulree Publications
First published 1994

All rights are reserved. Except for the purpose of review, criticism or private study no part of this publication may be reproduced, stored in a retrieval system, or transmitted, in any form or by any means, electronic, electrical, chemical, mechanical, optical, photocopying, recording or otherwise without the prior permission of Amulree Publications.

ISBN 0 9521126 2 0
The Velocette Saga (Hardback)

Contents

An Explanation	7
Forewords	8
Acknowledgements	9
Introduction	10
The early years	11
From 70 mph two-strokes to the 350 cc overhead-camshaft TT winner	17
The 1922 Denley TT mount	26
It would have been too "story book" if the cammy Velo had won in 1925	29
Early cammies	32
Alec Bennett's story of his TT rides for Velocette	37
"Spring-heeled Jack"	39
Michael Tomkinson	40
The KTT: for beauty of line it has no equal	44
KDT - Kammy Dirt Track	50
'Ugly Duckling' gearbox	51
Velocettes supercharged a KTT and Willis called her 'Whiffling Clara'	52
Willis-ese: A valve was a nail, a piston a cork, the test shop a din house...	54
The "dog-kennel" engines	58
No matching numbers!	60
The revered GTP. And the KTP	63
"Generally Tight Piston"	68
KTTs and the new M series	70
Interviewing Eugene Goodman	77
The MAC was more than just a "cooking" motor	79
The M Super Sports	81
Eugene would have been surprised: the David Holmes MOV/MAC racers	82
"A magnificent engine - but as for roadholding..."	86
Billy Wing; Velo agent	90
The Mark 6 KTT: mystery solved ?	94
When the black and gold beat the black and silver	97
Stuart Waycott and the ISDT 600	100
The Wilkinsons - a Velocette family	107
Veloce horses	110
Peter Goodman: boss's son who had riding talent	111
The racing stories of Ernie Thomas	117
Model O... The Secret Superbike	122
Roarer stands down and Germany takes the Senior TT	132
Roarer reborn	134
Arthur Lavington	141
LE - a motorcycle for Everyman	144
Birth of the LE	146
Irving's Light Engine design: more revolutionary than it appears	154
Arthur Taylor	156
Talking with Bertie Goodman	159
Director who likes his own bikes	160
What Four?	172
Ralph Seymour, of Thame	174

Post-war roadsters	179
The unmistakable exhaust note	183
First and last Thruxton	184
Putting Britain back in the records book	188
Jack Passant	190
Neil Kelly: Velocette's last TT victory	192
That clutch	194
The last all-new design from Hall Green	197
Matt Holder and the Velocette Motor Cycle Company	201
Velocette Days, by Bill Snelling	202
Whatever happened to Veloce Ltd?	208
The latest engineering Goodman	214
Numbers game	216
The survivors	218
Velocette Specials	220
The Velocette Owners Club	225
Club Spares Scheme	226
The LE Velo Club	227
Bibliography	228

An Explanation

The author in the paddock at Oulton Park on the 1930 Velo which looks like a KTT, but is not (the forks are KTT, though). For racing, the traditional exhaust pipe extending beyond the rear-wheel spindle is replaced by a shorter, more efficient but less aesthetically pleasing pipe.

This is not a normal book. It is the result of a growing (and flattering) demand from vintage and classic motorcycle enthusiasts all over the world for a reprint of "The Velocette Saga" which I wrote as a serial story for the monthly magazine *Motorcycle Sport* in the early 70s. Not a normal serial story either for it was not written first and then cut up into convenient monthly portions but was written 'live'. Each month, as spare time permitted, I went on a voyage of exploration into the fascinating past of the Goodman family and their Velocette motorcycles, never knowing from one month's instalment to the next who I would meet and what I would learn, for most of the key figures in the story were then alive. In a sense I was taking my readers along with me and many became involved and volunteered information. My good Vintage Club friend Ivan Rhodes became more involved than anyone and accompanied me on many of my expeditions. The first hand information he gained this way formed the foundation and the inspiration for his masterly book "Velocette, Technical Excellence Exemplified."

The "Velocette Saga" became a labour of love as it progressed, and an important part of my life — and, as I learned, was valued by a host of Velo fans, and even by non-believers only the other day a friend who I thought to be a life-long Velocette enthusiast confessed that he had not been until he read "The Saga". Once the stock of back copies ran out the demand for reprints started and although I was willing to co-operate in producing a book based on "The Saga", and actually wrote update material on a couple of occasions, commercial conditions were never favourable: and nothing happened. It took the enthusiasm of a dedicated Velocette rider to overcome the obstacles: Bill Snelling, this is your book! When of course I re-read "The Saga" it was clear that in two decades a lot had happened, including the liquidation of the firm, though enthusiasm for the marque had survived. So once again I went on my travels to meet the people and the machines to bring "The Velocette Saga" up to date. As before it is more about people than technicalities. In my book people matter most.

C.E. Allen, BEM Ibstock, 1994

Forewords

By Dr. Joseph W.E. Kelly, M.Sc Bradford, C.Eng. M.I. Mech E.

In 1994 I became a Pioneer member of the Association of Pioneer Motorcyclists, thus celebrating 50 years of motorcycling.

It has been 50 years of a fascinating and varied hobby. A period of trials and grass tracking undertaken with more enthusiasm than success. A magic carpet to explore my own country from Land's End to John o' Groats. Several years working my way round the world with an Ariel twin. Forty years of collecting, renovating and riding classic machinery, and five years of research into the economic history of Veloce Ltd.

Throughout the whole period I have been an avid reader of motorcycle literature, initially the weekly magazines and latterly the wealth of books that have become available. But no other scribe has had more influence on my motorcycling life than Titch Allen. In 1948 he wrote an article about the embryo Vintage Motor Cycle Club and his predilection for Brough Superiors. I have been a member ever since and have owned three Broughs! In 1967 he wrote about a Triumph TR5 Trophy. I have owned one ever since! At least I am not alone; it is hard to imagine anyone who has had more influence on the life of motorcyclists, when one considers the growth of the VMCC.

It was in the realm of Veloce research that Titch Allen was especially helpful, pointing one in the direction of useful contacts. The Velocette Saga, published in 'Motorcycle Sport' was a very valuable source of information. I am pleased that expanded, I gather, it is being published in book form. It is an interesting, factual and balanced account of the firm's history, paying as much attention to the Viceroy and the LE as it does to the racing models. It illustrates that fact that the impact of Veloce Ltd. was out of all proportion to the size of the firm; I am sure it will be a success.

J.W.E. Kelly

By J. R. Clew, Hon. Secretary, Association of Pioneer Motor Cyclists

I am honoured to have been asked to write this foreword to Titch Allen's book about Veloce and Velocettes. Titch was the first person to delve deeply into the story of one of the most elite of all the British motorcycle manufacturers, to give rise to a 13-part serialisation in *Motorcycle Sport*, commencing with the January 1969 issue. His "Velocette Saga" disclosed many previously unknown facts about a company that enjoyed a unique relationship with all its employees, regarding them as an extension of the founder's own family. Such an atmosphere is unlikely ever to be seen again, but it lives on in the machines they made.

When the first part of the Velocette Saga appeared in print, the late Bob Burgess and I were still researching the history of the company for our forthcoming book, Always in the Picture. I recall the appearance of this "rival" work caused Bob some concern initially, as he provided some of the information Titch used at a much earlier date. Fortunately his fears proved unfounded. If anything, the two separate works were complementary to each other, the Velocette Saga creating such an interest that it acted as a spring-board for the launch of our book. Now, at last, the Velocette Saga has become a book in its own right.

Often short of money, Veloce Limited produced some remarkable machines and enjoyed many competition successes - even their 24 hour record of 1961 still stands. Inevitably, the collapse of the British motorcycle industry caught up with them, hastened by the ill-timed diversion into the Viceroy scooter. But the legend lives on and Titch Allen's book will serve only to strengthen interest.

Jeff Clew, Sparkford, Somerset

Acknowledgements

The author acknowledges with gratitude the assistance he has received from the following, without whose help this book could not have been written.

Cyril Ayton, Editor of *Motorcycle Sport,* who published the original "Velocette Saga", not knowing what each instalment would provide and has been a constant support and encouragement ever since. The late Bob Burgess, former Service Manager at Hall Green, who gave his help unstintingly though at the time he was assisting Jeff Clew to write their Velocette history "Always in the Picture". Jeff himself went on to become a prolific writer and editor, yet was always willing to help a fellow vintage motorcycle enthusiast.

The Goodmans, Eugene, his son Peter and nephew Bertram and Simon, his son. The Denleys, Ethel and George, all so helpful, so patient, even when taken back to times they might have preferred to forget.

The designers: Charles Udall and Philip Irving, poles apart in personality but with a single purpose on the drawing board... to produce a better Velocette. Both patient under questioning about the whys and wherefores.

The riders: too many to list but let me thank Stanley Woods and Freddie Frith in memorium to represent all who rode Velocette in competition and told me their stories. All of them touched a little by the magic which flowed from Hall Green.

The dealers: mostly 'rider agents' in the beginning who had become personal friends of the Goodmans and therefore part of the Velocette 'family'. Loyal to the last they stuck with the firm through not much thick and a lot of thin, proud to have been on friendly terms with Mr. Percy, Mr. Eugene and later Mr. Bertram and Mr. Peter. It was quite Victorian. You may have heard of Archers of Aldershot and Tiffens of Carlisle, they being life time Velocette agents as far apart as I can think of, and such London names as Stevens, Orpin, Dodkin and Roy Smith. There were countless more and many helped me but none so much as Freddie Frith of Grimsby (and his right hand man, Val).

But it was smaller agents, never famous or wealthy and so long gone now as to be mostly forgotten, who really started me on my personal Velocette Saga; and countless other young 'sparks' too. Ironmans of Peterborough, Petty of Leicester, Wing of Nottingham, local agents I remember as yesterday. They rode Velocettes and convinced me that there was something magic in the name.

Back to the present, so many friends have helped over the years I cannot list them all. But the one who has helped the most over the longest time, long before The Saga began, is Ivan Rhodes. A close and staunch friend, he needs no introduction because he is President of the Velocette Owners Club, is a world authority on the marque and owns an important collection of factory racers — and is always ready to help with knowledge or parts; I have not referred to him in this book as Ivan the Generous for nothing.

I am proud that my 'love affair' with Velocette motorcycles led me to seek and meet so many wonderful, interesting people, yet saddened because I cannot go back and ask questions I should have asked; and nor in so many instances can anyone else. My words, once limited by time and now by space, must suffice.

Finally a list of those who have given specialist assistance for which I am deeply grateful. Pete Busby, Rod Coleman, Bob Higgs, Clive Hawkins, Dave Holmes, Dave Masters, Jonathan Wortley; and special thanks to David Matthew Scott Holder of the Velocette Motorcycle Co. for his permission to use the Velocette logo.

C. E. A.

Introduction

The title, The Velocette Saga, for a series first serialised in *Motorcycle Sport* in the late 60s and early 70s, was chosen after much deliberation. I did not wish it to be thought that I was cashing in on the success of a current entertainment phenomenon by using the word saga, but I felt the normal "story" was inadequate in the case of the Velocette motorcycle. In its simple dictionary sense, the word saga means no more than a tale — a Scandinavian legend — but by common usage it has come to be regarded as a record of outstanding achievement or the chronicle of an unusual family. Both these definitions fit the Velocette motorcycle and the Goodman family perfectly.

For over 60 years Velocette motorcycles had — indeed, still have — a rare, exclusive character all their own, a character which endeared them to two generations, to fathers and their sons (perhaps made a few enemies as well, for a strong personality makes both friends and enemies). That character stems from the simple fact that from the very beginning Velocettes were directed by one family — the remarkable family Goodman. With an unusual honesty of purpose zealously maintained through good times and bad, the Goodmans, father and sons ... and *their* sons too, and not forgetting a daughter... had one real object. To produce a motorcycle which would be a pleasure to ride.

That sounds a simple thing to do. But is it? All the time there is the niggling temptation to cheese pare here and there to make more profit, to jump on the trendy bandwagons of fashion to catch more sales. To dress up mutton as lamb. To make a few special racers for prestige purposes, but churn out bread and butter bikes for the peasants. The Goodmans scorned all these short cuts to commercial fortune, being content with the smaller market of knowledgeable enthusiasts who could appreciate quality rather than quantity in tinsel wrappings.

(Not until the 1960s did Velocettes have chromium-plated tanks and handlebars, and clothe the engine/gearbox in fibreglass panels; they preferred instead to polish their alloy castings to a mirror finish.)

Only once or twice, and then perhaps tongue in cheek, did they succumb to the demands of the sales department and produce a model which was more expedient than functional, and in each case the model was so short-lived that no real harm was done. For most of 60 years the Velocette remained uncompromisingly functional in sedate black and gold, but was on several occasions technically way ahead of its contemporaries. It was always stubbornly unconventional in certain small details, like the clutch and kick-starter mechanisms — trifling peccadillos which add to character and escape from conformity. Trifling indeed alongside the merits of smooth performance, long life and, above all, good handling with which each and every Velocette has been richly endowed.

No wonder, then, that over two generations Velocettes won not only world-wide esteem and enough competition successes to fill a book, but gained a unique following of dedicated enthusiasts.

The unique relationship between a man and his motorcycle, be it love or hate, is beyond logical explanation. A motorcycle, on scientific analysis, is merely a compound of ferrous and non-ferrous metals with a few plastics thrown in, yet in the hands of man it can (if it has character) be such a faithful servant and partner that a communion is established between the animate and the inanimate. A communion which evokes affection, loyalty, devotion and, in advanced cases, a determination to forswear all other makes. Thus is a one-make enthusiast born, and among the most loyal are those who have always defended the faith of the Velocette. And why not, for in truth they are honouring the creative work of a talented family and all the loyal workers who were inspired by their leadership to produce of their best in craftsmanship.

C. E. A.

The early years

The Goodmans were determined to make motorcycles which would be a pleasure to ride

Hall Green Works, York Road, Birmingham 28. Home of the Velocette from 1927

Because this is a family story a family tree, in so far as the Velocette history is concerned, may be helpful.

The roots of this family tree go right back into the 19th century when Johannes Gütgemann left Germany because he did not like the militaristic spirit growing there.

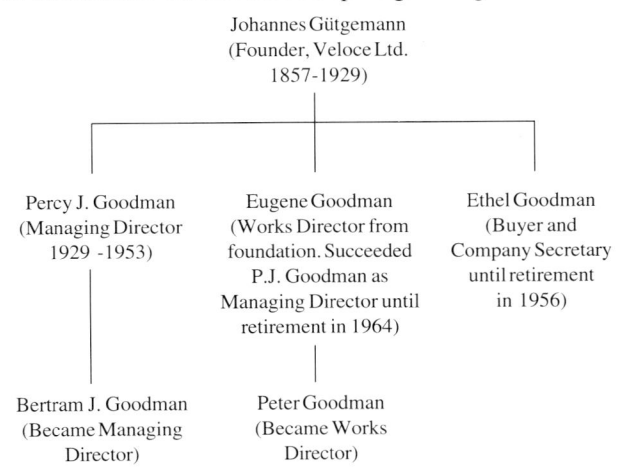

```
              Johannes Gütgemann
              (Founder, Veloce Ltd.
                  1857-1929)
    ┌─────────────────┼─────────────────┐
Percy J. Goodman   Eugene Goodman    Ethel Goodman
(Managing Director (Works Director from  (Buyer and
   1929 -1953)     foundation. Succeeded  Company Secretary
                   P.J. Goodman as      until retirement
                   Managing Director until   in 1956)
                   retirement in 1964)
       │                  │
Bertram J. Goodman   Peter Goodman
(Became Managing    (Became Works
    Director)          Director)
```

Johannes married Elizabeth Ore, the daughter of a Shropshire watchmaker and settled in Birmingham. His first business venture was the take-over of a patent medicine and pill firm known as Isaac Taylor and Co. Eugene Goodman in later years was wont to recall rolling pills as a small boy and say in jest that the family might have been better off if they had stuck to doing just that... bearing in mind the commercial success of Beechams, another pill firm.

By 1896 the bicycle boom was at full tilt and Johannes started another firm, Taylor Gue and Co., Ltd., engineers, to make the Hampton bicycle as well as cycle fittings for other manufacturers.

Taylor Gue prospered, and among their customers for frame parts were the makers of the Ormonde motorcycles, one of the better pioneer motorcycles, which employed a Belgian Kelecom engine. (Most of the pioneer machines had imported engines.) Paul Kelecom, designer of the engine (and subsequently designer of the FN four) joined up with the Ormondes to make a new Ormonde Kelecom, and Taylor Gue made the frame and parts of the engine. The joint venture soon floundered in a matter of months, in 1904, and Taylor Gue took over (could be they had not been paid for their work...).

In 1905 they produced their first machines and christened it the Veloce (in music: *with great rapidity*). A good name if a little extravagant for a rather undistinguished 402 cc inlet-over-side-exhaust-valve with direct belt drive. It does not appear to have caught on (only one has survived) and after only a few months the firm of Taylor Gue Ltd. folded. But the name Veloce was salvaged from the wreckage and Johannes started a new firm, Veloce Ltd., in 1905 which made cycle parts, roller skates, anything that anyone wanted... including rickshaw wheels. Son Percy had a spell out in India working on imported Wolseley cars and made such an impression that the importer persuaded him to return to England to design a motor car for export. The importer promised to take the entire production. Percy did design the car, under the company name of New Veloce Ltd., and a prototype was built with proprietary components. By this time war clouds were rolling and anyone of German extraction was liable to discrimination. It is not, therefore, surprising that the family adopted the English version of their name — Goodman. Previously, in connection with Taylor Gue Ltd., and Veloce Ltd., Johannes had been known as "Mr. Taylor".

The car project fell through but the prototype ran for years in Birmingham and was last seen with a van body, and the sons Percy and Eugene turned their energies to motorcycles.

In 1910 they began with a new and very sophisticated lightweight 276 cc four-stroke, as well as a simple 500 cc side-valve single with direct belt drive and the option of a hub gear, which was marketed as a VMC.

The first all-Veloce engine unit was early proof of the brilliance of the Goodmans as innovative designers. At a time when direct belt drive from engine shaft to rear wheel sans clutch or tensioning device was the norm, and oil dripped into the crankcase via an adjustable sight feed was considered adequate provision for lubrication, the Veloce had two speeds engaged at will by selective clutches operated by a rocking pedal foot change. It was of the then popular overhead inlet, side exhaust layout... this being perhaps the best layout at the time to ensure reliability of the exhaust valve. Little was known before the first world war of alloy steels capable of retaining their strength under near red heat conditions; it was war-time aircraft engine development which gave us better valves and made the ohv engine a reliable proposition.

Two features stand out and show the Veloce mastery of design. The first was the economy in layout and dovetailing of components which resulted in an engine, gear and magneto unit no larger than contemporary simple direct-drive engines. There were two spur gears side by side on the crankcase, of different dimensions. These engaged with two larger spur gears hollowed out to form the female part of a metal cone clutch, and loose on the countershaft, which lay in front of the crankshaft.

One of these gear wheels cum clutch bodies also drove the magneto via an idler gear, and the other gear carried the cams on its extended boss. In a word, the countershaft assembly was also the camshaft, and the belt pulley on the end in front of the engine was in an ideal position because it permitted a longer belt than normal and consequently more wrap round the pulley. The pulley was balanced to some extent by an outside flywheel on the offside of the unit and because it was an over-hung design - a feature passed on to the long line of two-strokes which followed - it was possible to balance it visually by the rocking pedal foot

The 1905 Veloce

With the help of brother Eugene, Percy Goodman designed and built a car under the name of New Veloce Ltd., based at the same Spring Hill factory as Veloce. It had a water-cooled 20 hp engine built by Johnson, Hurley and Martin, and a body by Startin of Aston. Although the car was used and demonstrated successfully, orders were not forthcoming and production went no further than the prototype which was last seen in van form before it was scrapped, George Denley driving it to the scrap yard in 1919

An advertising promotion for the 276 cc unit-construction Veloce

change mechanism.

Oil was contained in a car-type sump beneath the engine-gear unit and was circulated by a gear pump driven off the end of the magneto drive. After passing a control valve with an indicator rod which was taken up alongside the petrol tank to reassure the rider that the system was working, oil was pumped into the hollow countershaft cum camshaft, where it escaped to ensure that the gears and clutch ran constantly in an oil bath. A final touch was a drilling in one of the cams which flung oil at the right moment into a cup on the big end eye, the piston benefiting from the resultant splash. It was an extremely well-thought-out system, but short-lived because the war halted production, and afterwards Velocettes decided to concentrate on two-strokes for lightweights. It was rather unfortunate that the only engine of this type to run in the 1913 Isle of Man TT developed a serious leak in its lubrication system and smothered itself with oil. This might not have been so serious, but the oil got on the belt and the rider, Cyril Pullin, who was to win the Senior TT next year on a Rudge, had to stop so many times to wipe it clean and regain a drive that he finished 22nd... and last.

The TT engine was an enlarged version of the two-seven-six machine, the 68mm bore being increased to 76mm, giving a square bore/stroke ratio and a capacity of 344 cc The smaller engine came out at 276 cc which we would now regard as an odd size, but was a common one in those times when there was no established 250 cc class, and the rather vague description of 2½ hp covered anything appreciably under 300 cc. It is interesting to note as early evidence of a Veloce policy which was rigorously maintained in the years to come, that the firm immediately afterwards offered the 276 cc model to the public, uprating it in 1914 to a 293 cc (2¾ hp) model in both ladies' and gentlemans' versions.

Oh yes, there was a ladies' model. As early as this, adventurous females were taking to two wheels, their fruit salad hats anchored firmly with veils tied under the chin and

The first two-stroke. A 1913 Model A with direct belt drive from engine to rear wheel but automatic lubrication. Owned by Ivan Rhodes and still going strong

their voluminous skirts - the more daring with divided skirts - revealing a twinkle of ankles. To accommodate these fashions, the majority of manufacturers made specially dropped top tube models, not unlike the present day adaptable push bikes and provided shielding for the cylinder and external moving parts. The requirements of a ladies' model were so rigidly laid down by fashion of the day that all makes looked much of a muchness... even flat twins like Douglas and Brough... and the Veloce was no exception. It was not until the immediate post-first war period that Veloce Ltd. came up with a distinguishing feature which made their ladies' mounts easily identifiable, and that was the introduction of a cast-alloy, colander-shaped shield to encompass the cylinder head.

I regret to say that I have not been able to unearth much information about the performance and characteristics of this clever little unit-construction model, the first example of forward thinking of the Goodmans. All I have been able to trace by way of a road-test is a "brief impression" from 1912 wherein the tester seems astonished at the easy starting... it was paddled away, there being no kick-starter. A kick-starter was added in 1914. The bike was so docile that immediately afterwards the gear pedal could be pressed into the neutral position to allow the rider to tuck in his coat tails.

Once the rider was on the move bottom gear was unnecessary. The unit was silent running apart from a slight "sing" from the gear drive. Significantly, the handling was described as "very good", a description applicable to every subsequent Veloce and Velocette. If the unit model did not sell it was probably because it was a little too advanced and pricey for the utility market, and the Goodmans lost no time in producing a two-stroke lightweight for the lower end of

the price bracket. In this move they were not alone for most manufacturers saw the two-stroke as the answer to a low-cost runabout. The first Veloce two-stroke was a particularly compact and tidy unit of 206 cc with the magneto tucked close behind the crankcase out of harm's way when the majority of designers stuck it on a shelf projecting in front of the engine, where it collected road filth. Two features were outstanding and destined to be carried forward on all Veloce two-strokes up to 1926.

The first of them was the offset to the offside of the cylinder in relation to the centre line of the machine, thus balancing the weight of the flywheel on the nearside and permitting a one-piece crankshaft with an overhung big end. The design broke away from the convention of a crankshaft split up the centre line with its attendant problem of alignment and gas sealing at a joint liable to "work". The crankcase was cast in one piece, like a saucepan, and the

---

The Motor Cycle of to-day must have

PERFECT

MECHANICAL LUBRICATION

FREE ENGINE & TWO SPEEDS

The ONLY Motor Cycle combining these advantages in ONE UNIT is

The VELOCE

46 GNS. GENT.'S   **48 GNS. LADY'S**

THE VELOCETTE
$2\frac{1}{4}$ H.P. TWO-STROKE
MECHANICAL LUBRICATION
Belt Drive 25 G, Chain Drive 2-Speed 30 G.

DESCRIPTIVE BOOKLET FREE.
GRADUAL PAYMENTS.

VELOCE LIMITED, FLEET STREET, BIRMINGHAM.

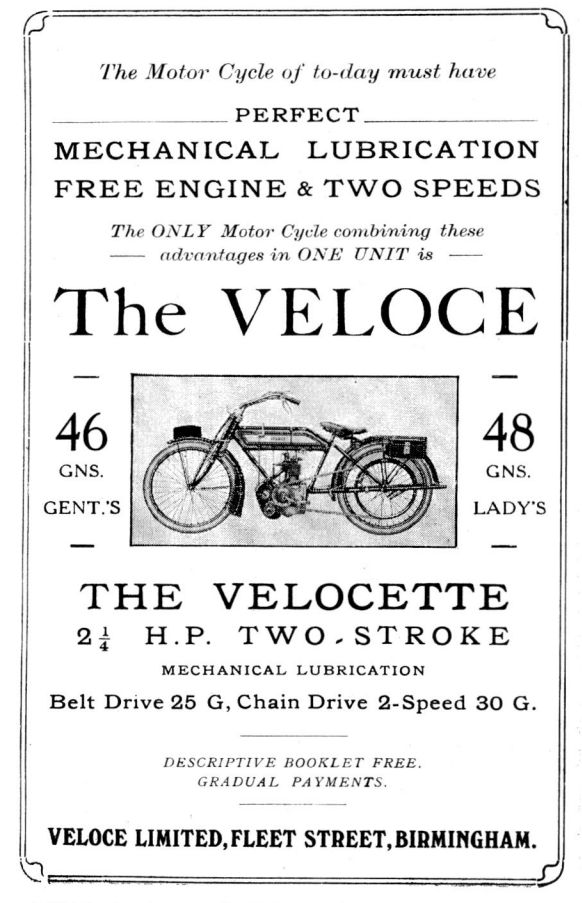

A 1914 advertisement for Veloce Ltd., offering both ioe Veloce and two-stroke models

internals assembled through the open mouth and finally sealed by a bolted-on lid.

The other important feature was the provision of automatic lubrication under the control of the rider. This was quite revolutionary at the time when petroil was regarded as God's gift to two-stroke designers and shows the Goodman's determination to do every job properly and avoid short cuts. The oiling system was ingenious and, although it was later to be changed in detail at various times, the basic principle was to used up to 1929... at that point the two-stroke range was completely re-designed and while the crankcase and crankshaft arrangement then succumbed to convention, the lubrication soon went one better, with the first throttle-controlled "posiforce" oiling system.

In the 1913 model the oil was contained in a compartment cast in with the crankcase and lifted by a ring conveyor (a system well known in stationary engine practice whereby a loose ring acts as a dredger to lift oil from a sump into a feed trough). In some descriptions of the first engines a chain conveyor is mentioned, but always the oil was lifted to a trough metered by a hand-adjustable needle valve and fed into the hollow crankshaft and so to the big end. Later in the year the ring conveyor seems to have been discarded in favour of pressurization by exhaust gases, and in 1914 a redesigned model employed induction depression to feed the oil.

Crankcase depression was also employed on a cut-price utility model in the 20s but all other post-war models had a proper oil pump to feed the oil. I mention all this to emphasize the Goodmans' preoccupation with the need for a mechanical lubrication which relieved the rider of more than a routine responsibility of filling the sump, and their distaste of the messy method of mixing oil and fuel.

The original 1913 two-stroke was offered in direct belt drive form at a competitive price of 25 guineas and in de luxe form, with two-speed countershaft gearbox, at 30 guineas... and all-chain drive. I am not prepared to argue about it, because so many odd-ball machines were produced in penny numbers in those days, but I firmly believe this was the first example of a lightweight single-cylinder two-stroke to be fitted with a countershaft gearbox, and all-chain drive. Scott devotees should note I have specified a single cylinder and a gearbox. And the novel feature of the gearbox, apart from the fact that it was made of cast iron and clamped on its platform by one bolt only, was that the final drive was outside the primary sprocket. Here is the root of the design feature which thereafter made every chain-drive Velocette different from all other machines. (With one notable exception - the 3-speed Scott - where a similar construction was later adopted for much the same reasons.)

The outboard final drive was clearly adopted because it incurred the minimum change in the layout of the machine from the basic belt-drive version and offered the twofold bonus of a quick change of gear ratio and an accessible rear chain, plus the later possibility of fitting oversize tyres. The offset of the crankcase to the offside had provided a narrow belt line with the minimum of overhang pull on the crankshaft. It would have been a pity to lose this sound engineering feature by outrigging the chain primary drive outside the flywheel to line up with a conventional outboard countershaft sprocket or clutch, so the final drive was arranged outside the primary.

The same reasoning applied when there was a demand for a clutch. To most designers, thinking of conventional

Velocette two-stroke of 1919: 229 mpg and "250-300 miles on one charge of oil". The nut on the crankcase of this early model retains a peg on which spun a slave crank web to assist with balancing the overhung driving crank. A trifle over engineered was the rack and pinion operation of the gears by the 'twist the knob' device at tank level. After all, there were only two speeds!

The Velocette Family Goodman rode the machines they built. Even the ladies. Mrs. Percy Goodman, mother of Bertie Goodman, rode this ladies model on holiday trips from Birmingham to Tenby before the first world war

clutches at least two inches thick with springs and operating mechanism, the addition of a clutch would have meant a major redesign of frame and engine unit. The Goodmans (I refer to them in the plural in this period of evolution because I visualise a team effort, though I am led to believe that the main inspiration came from Percy) got round this problem by evolving a clutch unique in its compactness - a wafer of a clutch compared with the conventional multi-layer sandwiches—and immortal for its mystery to the uninitiated.

But to get back to the 1913 model on which there was no clutch.... The gear change was by means of a Bowden cable and lever on the handlebar, a feature shared with the Baby Triumph two-stroke announced at the same time. All else was conventional in the lightweight field, if you accept the full-loop frame which, again, was to continue unchanged in principle until 1929. All-up weight was 112 lb for the standard and 112 lb for the De Luxe two-speed model. The Levisette, an ultra-lightweight version of the Levis two-stroke and rather crude in appearance by comparison, with its magneto stuck out at right angles to the large crankcase, weighed 80 lb.

Levis called their baby Levisette. Veloce called theirs Velocette. Both obvious diminutives. Who christened their babies first is anyone's guess and of no importance beyond the fact that one name was to stick out and become one of the best-known names in the world of motorcycles. Which brings me to an amusing story which by rights should come later but seems appropriate right now. The name Velocette was, of course, only intended for the lightweight models of Messrs. Veloce and had a slightly effeminate sound to it, which suited the two-strokes because they were dainty in appearance and charming in manner. When in 1925 the Goodmans launched Percy's secret weapon, the overhead-camshaft three-fifty, they naturally reverted to the more robust and masculine Veloce name. But by this time the Velocette two-strokes had built up such goodwill that when dealers found 'Veloce' on the tanks of the cammies they protested to a man. The firm saw the point and hurriedly switched the transfers to 'Velocette'.

Virtually identical to the 1914 model, this is a Model D1, but fitted with D2 forks, made in 1919 and beautifully restored by Velocette enthusiast and 24-hour race entrant Michael Tomkinson. Note tank-top handwheel control for the clutchless two-speed gearbox. The nut on the crankcase boss retains the pin on which a slave crankshaft bob-weight rotates. The machine can be seen in the Stanford Hall Museum

From 70 mph two-strokes to the 350 cc overhead-camshaft TT racers

Important milestones in Velocette progress

Apart from a capacity increase to 220 cc, detail improvements only were made to the Velocette two-stroke for 1914. A refinement was the use of a spring cush hub for the rear wheel, coil springs being set in an annular space between the brake drum and the sprocket. James used a similar idea at the time, too. It took some of the shock out of clutchless, wide-ratio gear changes but was soon abandoned. In fact, all the later two-strokes between the wars were without any form of transmission shock absorber and were none the worse for it. The other modification for 1914 was novel. To improve balance (and perhaps to improve crankcase compression) a slave bob-weight was driven by the overhung crankpin and run on a peg set in the outer crankcase door. The bobweight was only a loose push fit on the extended crankpin, but seems not to have given any trouble. It was pointed out at the time that this construction enabled the balance factor of a normal symmetrical crankshaft to be obtained without the problems of sealing the outer shaft. Having had some experience of early two-strokes relying on PB bushes for crankshaft support and sealing (wear in the bushes soon upset carburation), I think they had a point. The final improvement was substitution of a tank-side control for the two-speed gear in place of the handlebar lever. One might think the handlebar lever ready to hand was preferable but, again, experience of such a control persuades me otherwise.

The Baby Triumph of 1914-1923 had this scheme of a Bowden wire to pull the gear into top and a return spring to push it into bottom. Top-gear selection was satisfactory, but you had a wait until the spring felt strong enough to engage bottom, and when the cable became stiff, it seldom felt up to it. In 1919, Veloce Ltd. did the job in a real engineer's way... almost with the delicacy of a model engineer. A neat little housing on the gearbox cover contained a rack and pinion mechanism for operating a push-pull selector rod and a shaft ran up to tank-top height, nicely mounted in a bearing and topped by an alloy hand wheel. Gears were obtained by a twist to left or right. Any other designer would have had a bell crank at the bottom and a line-rod to a tank-side lever.

In 1915 the engine was altered with a new crankcase casting, plus a new head and barrel. When the same model was reintroduced in 1919 a rear drum brake had been added, but no sign of future plans, although in that year Percy Goodman patented his famous (old hands with experience of it right up to 1931 may prefer the word infamous) roller-action kick-starter mechanism. This worked on the principle of a roller on an inclined plane freewheel... a mechanism used with great success on car freewheels. It was on the far end of the layshaft remote from the outer cover, unlike most kick-starters, and avoided the trouble endemic in ratchet, pawl and spring kick starters. There were no pawls to break, teeth to shear and springs to fail.

Like other Velocette mechanisms, notably the clutch, the roller kick-starter was very good when it was very good... in good condition... but there was little margin for

Brake test on Brooklands Test Hill in the 1920 ACU Six Days. George Denley lowers his Velocette carefully with aid of the Goodman-designed internal front brake

Outstanding in design and performance, the twin-port, mechanically lubricated model G Velocette engine of 1923 was evolved from the 1922 TT engine. Tuned versions were capable of 70 mph, touring models of 150-200 mpg. In foreground is the crankcase "stuffer" and screw-in crankcase door. Big end was a plain bush

wear and it did not suffer fools or bodgers kindly.

For 1920 the two-strokes were given the "face-lift" of a raised top tube and wedge-shape tank, full duplex loop tubes and Brampton forks; this was the model D2. The "works" model sported an internal expanding front brake. This brake was designed by Percy Goodman and sold to Webbs, the fork manufacturers. It was the beginning of a long and successful business friendship between the two firms. Up to 1926 Velocette used Druid and Brampton forks, but eventually fitted Webbs exclusively. It was the "new look" 1920 model that brought the firm their first really significant competition success.

Previously, in 1919, Eugene Goodman had won a Gold in the ACU Six Days' Trial (an event which was designed to improve the breed of ordinary road machines rather than develop trials machines), but in 1920 a three-man team, Eugene Goodman, Stan Jones and George Denley, all won Golds and walked off with the team prize from under the noses of riders of machines of all capacities. This, in Denley's words, "put Velocette on the map". Next year the map extended to the Continent when Denley won a Gold in the Paris-Nice Trial and took first place in both the 250 and 350 cc class in the La Turbie hill-climb. And a French Velocette rider, M. Berger, pulled off the trials championship of France.

In the Isle of Man in 1921 could be seen some of the shape of things to come. Veloce fielded four two-strokes which were technically ahead of most rivals and were to set the pattern for a production range up to 1927. Gone was the old horizontal top tube and box-shaped tank, to be replaced by a nicely dropped tube to give a lower riding position and a stylish wedge tank. The engine followed the general construction principls of the original engine, being offset with an overhung crank, but was enlarged to a full 249 cc (63 x 80 mm) and had more generous finning. The 250 cc class of the Junior TT had been sponsored by *The Motor Cycle,* who had given a cup to persuade the ACU to encourage the development of lightweights. There had been much discussion as to whether the limit should be less than 250 cc; some two-stroke manufacturers, fearing the invasion of four-stroke two-fifties, had urged a lower figure... 225 cc was a popular capacity. Percy Goodman, however, came out strongly in favour of 250 cc, arguing that while 200-225 cc two-strokes might be satisfactory when new and in good tune, the general public might become disgusted with lack of performance later on.

At a period when belt-drive and clutch-less two-speed gears were common on racing machines, the 1921 Velocettes stood out from the rest. They had all-chain drive, a three-speed gearbox with clutch, and internal-expanding brakes on both wheels. Belt-rim brakes were considered good enough by most manufacturers. Two features which were tried out, and presumably found to be unnecessary, were enclosure of the flywheel and primary drive in a frying pan-type alloy cover, not unlike that used on Villiers flywheel magnetos, and torque stays linking the engine unit to the rear spindle. The team did quite well, W.G. Harrison finishing third behind the Levis star, Geoff Davison, who was beaten by a four-stroke, Doug Prentice's side-valve New Imperial JAP. George Denley, who we shall meet again later in the story (he married Miss Goodman and eventually became Sales Director), finished seventh, R. Humphreys fifth. Rex Judd, later to break records on Velocette two-strokes, retired.

George Denley might have finished higher but for two

misfortunes, one technical the other comical... at least he chuckled about it later as he told the story from his fireside. The first misfortune was the loss of his bottom gear which meant that he had to run alongside at various points on the Mountain. Not too much of a handicap for a young man who was rather on the lanky side - not until his braces broke! They did not break next year, you can bet he made quite sure of that, but alas, his frame did.

Detailed improvements and some innovations which were to be passed on to subsequent production models were seen when the 1922 race came round. The frame, less torque stays, was a little longer and the engine slung lower (incidentally, all Velocette two-strokes up to 1927 had cradle frames). Instead of the more common through bolts and engine plates, Velocettes preferred to clamp the crankcase to the cradle, and at one time the rear engine bolt was secured by taper cotters.

The biggest improvement was in the engine which now had an aluminium piston — most two-strokes still had cast-iron ones — and an aluminium, detachable cylinder head of beehive shape and two exhaust ports well splayed and with separate pipes to the coffee pot silencer. The offset of the engine in the frame precluded symmetrically exhaust pipe stubs, and the head-on appearance was a trifle peculiar, but one that came to command respect.

And, another important milestone in Velocette progress, there was a neat little mechanical reciprocating rotary pump built into the crankcase and driven by the idler gear of the magneto drive. Oil was pumped into the crankshaft via a port in the long PB main bearing bush and found its

Honora Gale rode with Ethel Goodman in early trials and added a touch of glamour to early publicity pictures. She became Bertie Goodman's aunt. Her mount is a 1920 model

Trials success put Velocettes "on the map". The winning team in the 1920 ACU Six Days Trial. From left: George Denley, Stan Jones, Eugene Goodman

way to the big end. The feed was controlled by a knurled knob accessible on the top of the pump plunger and as this rose and fell with the reciprocation of the pump plunger, the operation could be checked visibly and by feel. From an efficiency standpoint the twin exhaust ports were the most important change, and of this Bob Burgess wrote:

"The Velocette engine was originally fitted with one exhaust pipe leading from two rectangular ports in the cylinder wall that merged into the main port. It was found that the 'bridge' across the port, which was necessary to prevent the piston rings trapping in it, became far too hot, and P.J. Goodman designed a cylinder with two widely spaced exhaust ports each sufficiently narrow to avoid trapping the rings, but of total area much greater than could be obtained from a single unbridged port, and was able to fin the outer portion of the cylinder between them and so improve the cooling. These ports were of irregular shape, roughly triangular in outline, with the 'Hypotenuses' facing each other. Inlet ports were of what can best be described as flattened V in outline and there were two openings from the transfer port into the cylinder wall.

"A special experimental engine was used in the 1922 Lightweight TT. This engine, which was never marketed, had a detachable aluminium-alloy cylinder head, and was the fore-runner of the G model of 1923; but the production engines used one-piece cast-iron cylinders and iron pistons."

One of these two-port experimental engines brought S.J. Jones home third in the Lightweight in 1922, but two other entrants retired. Geoff Davison won on a Levis, staking all on light weight... himself in shoes, cord breeches and a sweater and his machine pared down to the bone... but an ohv Rex Acme Blackburne was second and it was the last time a 250 cc two-stroke was to appear in the first three for many years.

The production models G which followed were pretty accurate replicas of the TT mounts, apart from the detachable head, and soon established an enviable reputation. It would be no exaggeration to say that they were the "Rolls-Royce" of small two-strokes, and their excellence and discreet black and gold finish gained them a following of connoisseurs. An exclusive kind of customer unimpressed by glitter and flamboyance who was to remain loyal to the marque for 60 years.

At various times in the early 20s Velocettes built a few hot model G twin ports for their rider-agents and many of these monopolized the quarter-mile hill-climbs (often on public roads) which were a regular feature of local club life at that time. They were usually tuned editions of the production machines but a few 'beehive models' based on the TT models were made and a number sent to Spain. A machine of this type, probably the model known as the HSS appeared in early vintage club events.

Early catalogues contain pages of small print listing successes in club events by private owners of Velocette two-strokes, and for a time in 1922 and 1923 two-fifty classes in speed trials were dominated by specially tuned models ridden notably by Harold Petty, Alec Bowerman and John Sylvester. Such demonstrations of the sporting capabilities of the Velocette two-stroke when most two-

"Rolls Royce" of vintage two-strokes. The 1924 model H. Note Brampton forks with spring top links to give "two-way stretch"

strokes were distinctly pedestrian were very good for sales.

But the best bit of publicity probably resulted from the remarkable performance of a woman rider, Mrs. Jennison, of Grimsby. Wife of the local Velocette agent, she set up something of a record by winning the 250 cc flying start class at the ACU Open Speed Trial at Grimsthorpe Park in 1923 at 69.9 mph. Her average speed over the course implied a maximum of over 70 mph, and locally she became famous as the first woman rider to exceed 70 mph. There can be no doubt that she had one of the "specially tuned" machines, but, tuning apart, they were all perfectly standard model Gs. Sylvester also took the standing-start award.

There's a delightful little story about Mrs. Jennison's 70 mph run which can now be told. It so happened that Miss Ethel Goodman and her riding companion, Honora Gale, had been to the North to ride in a trial and took in the Grimsthorpe Park meeting on the way back. Mrs. Jennison had done one run but wanted to do better. Who better to ask for advice on carburettor settings than *the* Miss Goodman, sister of the designer? Miss Goodman never professed to be at all technical but she had a good memory. She remembered typing out a list of carburettor settings for speed work for brother Percy and was able to repeat it, "parrot fashion" she assured me, without knowing what it really meant. Miss Jennison followed the instructions — and 'bingo', the 70 mph run!

Miss Goodman had flown the Velocette flag bravely in 1921 when with Miss Gale (later to be aunt to Bertram Goodman, later Managing Director) she rode in a ladies' trial in France which ended with special tests (the course marked with skittles) at the Parc des Princes cycle track in Paris. She was not the outright winner — a broken fork spring caused by a bumpy level crossing and the miles of pavé cramped her style — but she did bring home a trophy. She recalled one awkward moment when she had to go on the wrong side of a motorist. "I gave him a wink as I went by". A gesture which I am sure would be appreciated by a gallant Frenchman.

The same weekend George Denley was making history in another way. He was the first soloist to make a clean climb of Screw Hill in North Wales, a fearsome climb which had defeated many experts and rejoiced in the title of "The Worst Hill in Wales". Later he was to make more history, being the first winner of the Travers Trophy Trial run by the Newcastle Club... a 200 guineas trophy, £50 in cash and a gold medal. Nice work, and more good publicity for Velocette.

Customer reaction to the Velocette two-stroke in the early 20s was very good. I could fill a book, this book, with enthusiastic testimonials from riders — and Velocettes did a book containing 100 or more letters from owners grouped under the headings of speed... most claimed 50-55 mph, one 65 mph; strength... tales of trials and touring in mountainous country and a tribute from a 14 stone rider; design... "has been really and honestly designed" and "designed, not merely assembled" were two tributes; workmanship... "the Rolls-Royce of two-strokes" was the most apt comment; flexibility... one owner said he had restarted on a 4½-to-1 gradient, another that it would run as slowly as a push-bike. Another wrote endearingly, "I am quite in love with the little monkey". Liveliness... one owner called it a little speed demon; economy... 150-200 mpg were common claims with an oil consumption of 2,000 mpg; power... many owners, it seemed, pulled sidecars or took pillion passengers and all claimed that no hill would stop them; reliability... one claim was 44,000 miles without trouble, another

Brooklands star Rex Judd pictured on his 1921 TT two-stroke. Judd set new world's records on a similar machine at Brooklands in 1921, capturing the British mile record at 70.31 mph

30,000. One owner wired for a spare on Wednesday midday, received it on Thursday morning!

The sophistication of the model G Velocette with its all-chain drive, internal brakes and twin-port engine with mechanical lubrication put it outside the price range of all but the well-off motorcyclist. Bob Burgess in his memoirs recalled that as an apprentice at the Austin works he could not run to the G model and had to make do with a model B, a cheaper single-port model (later, when he got on good terms with Eugene Goodman, he was able to buy a model G two-port barrel and piston and "hot-up" his machine). The model B was one of a number of cheaper models described as light models and designated A, Ac and B and aimed at the lower range of the market, which was being overrun by assembled machines from small manufacturers using proprietary engines, usually Villiers, and direct belt drive or bought-in gearboxes. Velocettes would only compromise so far in the interests of expediency and with the exception of a special cut-price model with final belt drive in 1923 and a short-lived model 32 — it sold at a new low of £32 — they retained the salient features of chain drive and internal brakes. Cost cutting came from the use of single down tube frames (in place of the characteristic twin tubes side by side) and the continued use of what was really the 1921-type single-port engine.

They were refreshingly frank about the economizing in the catalogues. "The engine. In outward appearance it resembles our H model (the model H was a direct successor to the model G). The chief difference in construction is its system of lubrication. The pump lubrication so successful in our model H had to be abandoned on account of its costliness. We had to fall back on our old and well-tried system, ie AUTOMATIC LUBRICATION by piston suction first used in 1913. Many engines of this type are running perfectly today." Contrast this with any present-day catalogue. No mind-bending image building ad. men then! The Goodmans wrote their own catalogues. "Percy and I spent the whole of one Whit holiday writing a catalogue," Mrs. Denley (Ethel Goodman) recalled.

The main difference between the model G and the model H which succeeded it in 1924 lay in the cycle parts. The frame was simplified and strengthened. Instead of a bolted-up chain stays with a malleable forging to support the gearbox, the chain stays were taken direct to the rear engine mounting lug running parallel to form guides for the gearbox which was grooved to accept them and clamped by a similarly grooved clamping plate. This gearbox mounting was used on all subsequent two-stroke models right up to the 1939 GTP. A larger front brake was fitted. The rear-brake anchorage was altered from a projection on the brake plate sliding in a guide in the fork end to a bolt in a slot in the fork end. A detail feature which was carried on for many years.

The model H never achieved as much fame as its predecessor, for the two-stroke was rapidly being overtaken and passed by four-strokes, but it remained a firm favourite with lightweight connoisseurs for a year or two. Although by 1927 the firm was concentrating on the ohc models, the old overhung two-stroke was not forgotten. It was given another face-lift in the form of a shapely saddle tank and long, "straggly" (it is Bob Burgess's word and I can think of no better) twin exhaust pipes and small tubular silencers on the ends. This was the model U; for utility, presumably, for that is what two-strokes had become in the middle 20s scene.

I am pretty sure that the "with it" exhaust system by no means improved the performance. As evidence of the role the model U was intended to play it reverted to the "cheap" suction oiling method. It was, however, the first production Velocette two-stroke to have a roller big end - 12 uncaged rollers. Early models, I am told, had standard H cylinders. Later models had less finning. There followed another face-lift job next year which was nearing the ridiculous after the sublime of the model H. This was the model USS which, being translated, literally meant Utility Super Sports. This had a bigger saddle tank, (the same size as those on the ohc

Tony Webb with the ex-George Denley TT two-stroke

model) and tail-end silencers. From secondhand experience of these models I would say that they were mutton dressed as lamb.

But the USS had one good point in its favour. It had a new cylinder barrel with a detachable head and the next year these components were to appear on a redesigned bottom end and provide a much better machine, which was the GTP of countless pleasant memories. The GTP is really another chapter, but as the USS pioneered its cylinder arrangement it is opportune to mention an unusual feature of the port design which was peculiar to all Velocette two-strokes from the G onward. Instead of using basically rectangular inlet and exhaust ports, like other manufacturers, Percy Goodman preferred a triangular shape. On the model G and H the twin exhaust ports were triangles with the sloping sides facing each other which gave more cooled metal between them. On the USS and GTP the triangles were reversed with the sloping sides outwards, giving less cooled metal — or less metal to overheat, this always being the "hot spot" with two-strokes. I am told that Percy preferred triangular ports for reasons of silence; they reduced the sharp crack of the exhaust and, used on the inlet port, reduced carburettor roar and probably blow-back as well.

I have devoted much space to the Velocette two-strokes of the early 20s for they were soon overshadowed by the ohc models. The four-strokes took the glory in the years to come, and still do in the minds of vintage enthusiasts, yet the two-strokes deserve their place in history almost as much, for they laid the foundation on which the Velocette cult was built and were in their heyday every bit as outstanding in design and construction as were the later four-strokes. I will add one personal memory. When I was a schoolboy each day was enlivened by the sight and sound of a three-speed twin-port which, with a box sidecar, was used to deliver newspapers around the villages. It could be heard miles away and it was a daily treat to listen for its rising snarl through the gears and its musical diminuendo as it faded into the distance. It was many years before I could afford a Velocette two-stroke. They always commanded a high secondhand value, which was of course a direct reflection of their worth, but own one I eventually did and it was as delightful as the music it made.

Regrettably, few G and H models remain to show their paces in vintage circles. This in itself is a tribute to their worth for it indicates that they "died with their boots on"... were ridden until they were completely worn out, not discarded early in life to be unearthed later by vintage collectors.

In the racing sphere four-strokes with overhead valves were developing fast, and largely as a result of the TT were gaining stamina as well as speed.

Velocettes could see the writing on the wall as well as anyone and, although Percy had ideas in another direction which I shall soon disclose, he and Eugene tried out some pretty weird experiments in an attempt to achieve parity between two-strokes and four-strokes. One was a flat-top piston engine with the combustion chamber compressed to a small cavity in the otherwise flat cylinder head. This was perhaps the first real squish cylinder head, and the secondary compression was astronomical. So was the power unit until it overheated and for the first time the overhung crank was found to be inadequate. The crankshaft was shrunk into

Velocette policy was always to use the Isle of Man as a test bed for new designs. The lessons learnt could usually be seen in next year's production models. The engine of this model, ridden by Fred Povey in the 1924 Lightweight TT, did not go into production, however. It had a poppet inlet valve in the crankcase instead of a normal cylinder port.

the crank cheek - the shaft pushed into a red-hot cheek - and it was thought to be immovable. But the flat-top engine contrived to turn the shaft in the crank and upset the functioning of the oiling system.

Even more way out was the supercharged two-stroke. This had a pumping cylinder sloping rearward behind the power cylinder and driven by its own crankshaft geared to the main crankshaft. The pump forced gas into what would normally be the transfer port via a small (according to Burgess, who remembered seeing it lying about years afterwards) diameter pipe.

An even more remarkable feature of the supercharged two-stroke Velocette (pumping cylinders usually driven by a slave connecting rod on the main crankshaft were tried by other designers) was that petrol injection was incorporated... timed injection by a piston pump driven by a cam formed on the pumping cylinder crankshaft. Surely a first for Velocette, and about 40 years before its time. Alas, they never got the supercharged engine running properly, and if they had, the metals available at the time would have not withstood the heat and stress.

Probably because of these abortive experiments they gave the 1923 TT a miss, but came back in 1924 with another experimental engine which did show real promise. This was the side-valve two-stroke, as Bob Currie termed it. I will rely on Bob Burgess' description:

"In order to compete with some chance of success against the racing four-strokes much more power was needed. To obtain a longer period of inlet opening, therefore, as one means of getting it, Veloce built a two-stroke 250 with a poppet inlet valve with which a much longer period of inlet opening could be obtained than was possible with a piston-controlled port.

"The cylinder and head were little changed from the earlier experimental engine, but the carburettor fed directly into the righthand crankcase, in which the valve seated. The crankcase was in two halves and the crankshaft was of the orthodox type with two webs and bob-weights and ran in two bushes. It was of built-up type with a roller-bearing big end. Lightweight alloy packings filled the spaces between the webs and the bob-weights. Oil was carried in a tank on the seat tube of the frame and was fed by a small pump fitted to the magneto chain cover on the right side of the engine and fed oil to the front of the cylinder between the exhaust ports, and to the inlet valve. The valve was operated by a cam on the righthand side mainshaft through a square section tappet. The valve spring was above the valve and rested on the flat head. In the interest of lightness, as the valve opened and closed at every revolution of the crankshaft, a square tappet was chosen in preference to a mushroom-headed one, and this was bored out to receive the head of the valve stem.

"Fred Povey rode the machine in the 1924 Lightweight TT but a mechanical failure of the contact-breaker pivot so reduced engine power, although not stopping it altogether, that he was unplaced.

"For reasons that cannot be ascertained now, the 1925 Lightweight entry, ridden by the late Phil Pike, used a normal inlet ported cylinder."

Since I had the good fortune to meet Fred Povey for a fireside chat I learned a little more about the poppet valve two-stroke and came to the conclusion that, valve apart, this engine does not so much represent the final development of the old overhung crank two-strokes, but is the true ancestor

Above: George Denley on a ladies model - either EL2 or EL3 - in the Travers Trophy Trial, trophy value 200 guineas, which he won

of the long and dearly loved line of GTP models which began in 1930. For the first time in Velocette history a built-up two-piece crank with a roller big end and mains was used, and the crank cheeks were built up into full discs by means of aluminium filling blocks. It is clear that Velocettes well understood the importance of high crankcase compression. It will be recalled that the 1914 type had a slave crank which served this end as well as providing more balance mass, and the redesigned post-war engines had a circular aluminium packing disc under the 'lid' of the saucepan crankcase. Gone was the old oil container in the crankcase, a separate tank being fitted on the seat tube, and the oil pump was mounted externally on the magneto chain case, now on the offside. Oil was fed to the inlet valve which ensured that it found its way into the crankcase and another feed went to the critical spot between the two exhaust ports.

This engine was capable of over 70 mph yet docile enough to be ridden on the road and in trials.

One little detail about its TT debacle — it was 10th and last of the 250 cc category — I can put straight.

Previous historians, probably perpetuating the same mistake, have always said that the pivot pin of the magneto contact breaker rocker arm broke and, although the engine continued to run, it lost most of its power. Fred Povey told me that it was the rocker arm, not the pivot of the EIC magneto which broke. A freak failure never heard of before or since but one that stopped the motor forthwith.

"I was lying about third, I think, with a lap to go when the rocker arm broke. It did not take me long to find the trouble, and I had a spare contact breaker, but it took me 25 minutes to fit it and get going. You see, I had my hands bandaged up because of the vibration of the engine. It could

The packing department, with a consignment of 1929 Model U two-strokes awaiting despatch

certainly go, that engine... it would do about 75 but tended to get too hot so you had to go fast down hill and take it easy uphill,".

One can easily imagine, if one has had one's hands vibro massaged into numbness by high-frequency vibration, the difficulty of changing a contact breaker with your fingers all thumbs.

Povey went on to tell me that it was the poppet valve engine which Rex Judd used at Brooklands to take short-distance records at over 70 mph and cleared up a mystery that had bothered me for many years. Long ago I remember Rex Judd telling me, apropos these record sorties on Velocette two-strokes, that he once rode one with two carburettors - one each side. I quizzed many people about a twin-carburettor Velocette two-stroke without getting a clue, but Fred Povey remembered it. At one time, he said, they tried an extra carburettor on the 'side-valve job', feeding through a normal cylinder port. Rex Judd was then, of course, a jockey riding under orders for the O'Donovan stable which was chiefly contracted to Norton and would not necessarily know of any technical details of his mounts.

But his memory was correct.

Study of contemporary Island shots of Povey's mount show that instead of the normal twin exhaust pipes into a 'coffee pot' expansion chamber in front of the engine there was a quite advanced expansion chamber system not unlike present-day two-stroke practice. On leaving the ports, the pipes swelled out like the famous "Oxford Bags" pipes of the 172 cc racing Villiers engine, then merged into a flattish expansion box under the engine which terminated in two small tail pipes with fishtails. Povey put up the fourth-best time at the end of practice with a lap in 43 minutes 43 seconds. All the machines ahead of him were powered by ohv Blackburnes.

This was not the last of this astonishing little engine. Phil Pike rode it in the Lightweight next year, but on this occasion the poppet valve was not used and an ordinary barrel was substituted. He retired. After that the story becomes a little bizarre. The engine, or the machine, was loaned to Arthur Cope, of the well-known Birmingham motorcycle dealers, always close friends of the Goodmans, who used it in trials. It passed to Frank Cope, who used it

The 1992 Denley TT mount

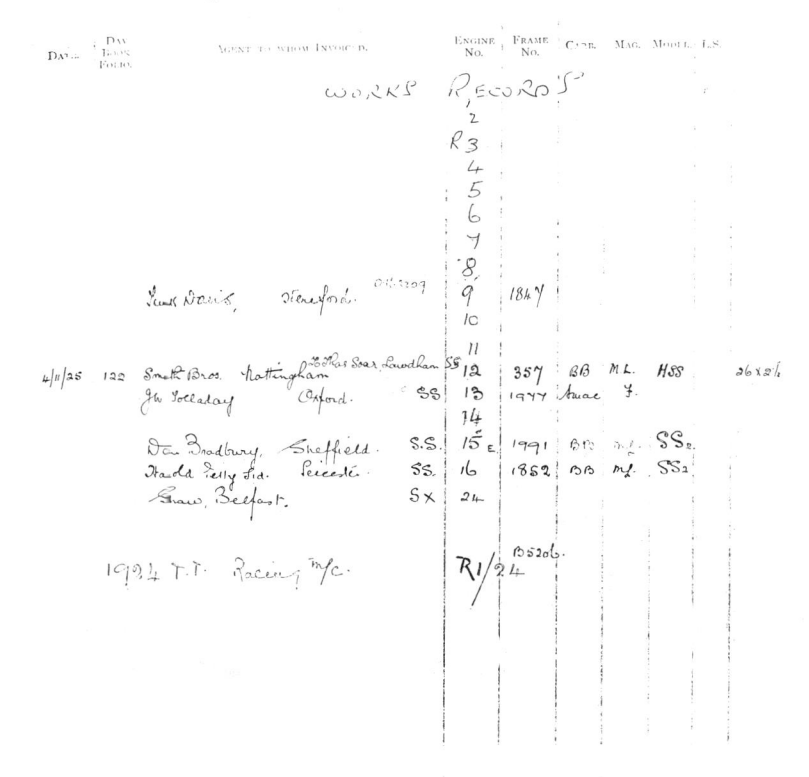

THIS COPY of the first factory ledger dealing with racing and competition machines shows how incomplete the records were. When I jokingly chided her about it, Mrs. Denley recalled that she always had difficulty keeping track of racing and competition models in the early days. "The boys (Percy and Eugene) used to loan machines to people and give away machines and parts to riders in competitions and never told me where they had gone". The entries indicate that only when a machine was sold through a dealer was an invoice issued and the blanks refer to engines retained by the works. Engine SS14 was fitted to the Denley machine.

Fortunately one of those 'works' machines survived and has been restored by Velocette two stroke enthusiast Tony Webb. His researches established it is the bike ridden by George Denley in the 1922 TT.

26

in trials with a sidecar and, after watercooling it, eventually fitted it into the boot of the BSA three-wheeler he used for long distance trials. There it did yeoman service as a booster engine, being coupled to the back wheel for steep observed hills. I remember seeing it perform, in a Land's End Trial, I think it was. The boot lid was propped open to provide ventilation and let out the smoke as the two-stroke lent its shoulder to the back wheel while the front wheels spun. Quite a sight.

There might have been more two-stroke development had not Percy been troubled with a recurring dream of a really advanced four-stroke engine which would bring the firm fame and success and, being a four-stroke, would be less liable to tantrums than the temperamental and unpredictable two-stroke. An engine that would combine the best features of some of the most advanced designs which had appeared in the automotive world. A four-stroke with the valves in a hemispherical head operated by an overhead camshaft.

For a start Percy set up his own private experimental shop in his private house. And it was there he was assisted by a dedicated youngster already old in experience who was to be his loyal Man Friday for many years. The youngster was Fred Povey, later, after some unsuccessful forays in the Isle of Man, to become one of the country's leading trials stars.

He came to Veloce Ltd. by a roundabout route via the James firm and Nortons; it may be that fate intended it so. He had started, a mere lad, as a tester at James, getting in a few minor successes at trials on the side — then moved to Nortons who were already among the elite of manufacturers. Nortons "posted" him to their outstation at Brooklands, where O'Donovan put the finishing touches to the BRS and BS single-gear side-valve hog buses. These were sold to the public with certificates of speed after O'Donovan had breathed on them. He took the engine, fitted it into a track machine, ran it in, and persuaded it to do the guaranteed speed. Povey's job was to run them in round the track. The job had its perks — such as the loan of a Norton for flying visits home at weekends. Around two hours, he reckoned, from Weybridge to Birmingham. One or two of his pals got wind of these trips and used to waylay him at Stratford for

---

Denley, who was Sales Manager of the firm, and later married Miss Ethel Goodman, partnered S.J. Jones, who finished third, and Rex Judd in the works team. Both Denley and Judd retired. Subsequently Denley's machine seems to have been sold to a Velocette agent in the Malton area of N. Yorks who used it in local sprints and hill climbs. Around 1928 it was rescued from a scrap heap by the Langton brothers of Leeds who did not appreciate that it was somewhat special. It lay with a number of other old machines until on a visit in the sixties I noticed a distinctive cylinder head.

When I told him that it was a special racing engine Oliver Langton said "If you know what it is you had better have it" and gave it to me. Many parts were missing but over the next few years I collected suitable replacements or made missing bits. Not being a machine suitable for vintage road racing it was rather low priority and I did not complete the restoration. I remembered that Tony Webb was keen on the early two-strokes and offered it to him as a kit if he would build it up. It was a wise move for in a short time he had it up and running. It has a perfectly shattering exhaust note which is how the "Beehive" engines were remembered.

George Denley, 1922 TT

1921 Model D, owned by Tony Webb

a burn up. One of them was a chap named Harold Willis.

But the job paled after a while because there was no future in it and Povey, having had a taste of trials, wanted more. His desire to get back to Birmingham was crystallized by an occurrence at Brooklands which, although it may startle the younger reader, does not surprise me one little bit because I have learned of similar pranks. It seems that one day when Povey had been circling the track all morning to run-in an engine which was intended for a record-breaking attempt he left it against the club-house while he popped in for a quick lunch.

When he came out — no bike. It turned out that the Brooklands practical jokers, of whom there were many and usually led by Freddie Dixon, had sneaked the bike away, and of all things, wangled it into the big oven in the restaurant kitchen. They "hotted it up, all right - part of it melted". This jape was not really directed against Povey but against O'Donovan, who was not the most popular of men, but, of course, Povey got the blame for leaving the bike unattended.

In his early days with James, Povey had met George Denley, who had been riding in the same trials, and Denley, now at Veloce, introduced Povey to Percy Goodman. Jobs were not easy to find in those days, especially the interesting jobs, and contacts were invaluable.

Povey said he often worked for Percy Goodman on the hush hush project in the private house in his spare time. The basic idea was straight forward. The ohc design was, he says, inspired by an early Rolls-Royce design: the flywheels with parallel crankpin and shafts and a web linking the mainshaft and crankpin bosses were based on the current AJS design which was fast becoming a leader in its class.

Percy Goodman's philosophy was, says Povey, "Only copy the best."

But it was a long job designing and making the parts for a completely new design with few machine tools and little money. There was another reason, apart from the shortage of funds. The pre-war design of the two-stroke, the only Velocette product, was getting out of date and was being eclipsed by the flood of rival two-strokes. Something drastic had to be done to win back a position in the field. There followed a period of intense development spurred by the desire to put up a good show in the Lightweight TT. This period, as we have seen, achieved limited success by bringing the near-standard three-port overhung two-stroke to the limit of its potential, but the failure of the way-out experimental engine to achieve any worthwhile results convinced the Goodman brothers that to compete with the rapidly improving four-stroke lightweights a two-stroke would have to be so complicated, with mechanical valves or superchargers, that the traditional advantages of the type in simplicity and low cost would be lost.

Percy was even more convinced that there was a better way to success — and success in those days meant a good enough show in the TT to promote the next year's production machines. Competition successes of any kind were valuable in this connection and Povey began to build up his own reputation as a trials rider on the two strokes — sometimes on the poppet-valve model he rode in the TT.

But while this "final fling" two-stroke development was going on, Percy Goodman persevered with his overhead-camshaft design, and gradually bits and pieces were made at the Veloce works until one or two prototypes could be completed.

It would have been too 'story book' if the cammy Velo had won, first time out, in 1925
Miracles do not (usually) happen in the world of wheels

The first ohc Veloce. This is the prototype model with constant-loss lubrication, the oil being fed to the cam box by the two-stroke pump on the end. Oil then found its way by gravity to the bottom end via the vertical drive shaft. A not very satisfactory system which was changed to the full pressure dry sump system which set a new standard in motorcycle design and made Velocettes so reliable

The original Velocette ohc design, as is well known to historians, differed in a number of details from the production layout. The lubrication system was crude, for one thing, and was soon found wanting. It was on the constant-loss system with a typically two-stroke Velocette pump fitted on the end of the cam box. Oil was fed into the cam box and then drained down the vertical shaft tube to feed the crankcase. Knowing how difficult it is to keep oil in a cam box with exposed rockers, I doubt whether much oil got to its proper destination. Another difference was that the vertical shaft was splined into the bevels to allow for expansion. Why this idea was abandoned in favour of Oldham couplings is not known - Nortons ended up using this scheme - but it may been dropped on the Velocette to provide a longer bearing for the bevels. But otherwise the prototypes were outwardly similar to the later production models. For the sake of convenience the three-speed box and single plate cork clutch of the two-stroke was used - this necessitating the unusually narrow "slimline" crankcase and unusually narrow chain line which became standard Velocette practice. Many have argued, myself included, that much of the success of the ohc is due to this feature. And have voiced admiration of the genius of Percy Goodman in designing a crankcase assembly so slim and rigid when others made them wider and wider until at one time there was a fashion for outrigger bearings to withstand the pull of an outboard chain. Alas for such misconceptions, such treasured illusions. It had to be that way to fit in with the transmission layout which went back to 1913! Whatever the reason, though, it turned out to be a good thing for it kept the Velocette dainty and slim while its rivals became ugly and overweight. I leave it to the technical minds and the Velo enthusiasts to fight out the question of whether the ability of ohc Velocettes to "buzz" — to rev sweetly and smoothly with less vibration than most others — is in part or wholly due to this slimline feature which places the main bearings so nearly under the centre of explosion pressure. For good measure — and to call for another round — they can discuss the theory, long held by Fred Povey, that the traditional excellence of Velocette steering is in part due to the "heavy bits", to wit, the flywheels, being nearer the centre line of the machine than most, as well as insistence that the major mass of engine and gearbox be below a line drawn through the wheel centres.

It would have been just too much like a story book plot if the cammy Velocette had won in 1925: the dedicated designer, slaving in a back room long into the night with his faithful young assistant, triumphing over factory might and all that. Such miracles do not happen in the world of wheels. Howard Davies came near to the schoolboy dream when he won the TT on his own make of machine, but he was, after all, using a well-tried engine tuned by the master tuner and the rest of the machine was out of the box of best mixed. The whole of the Velocette was new... the bits culled from the two-stroke range were new to this power and this speed and there just was not enough time to test and temper the model for such a gruelling. They tried hard to sort out all the snags that cropped up in practice.

First production K model, in 1925 — in sports trim

K — touring style

"The first week of practice we nearly lived at the Athol Garage (well known for last-minute TT mods.). The steering head races (cup and ball like the two-strokes) broke up, the cast-iron brake drums distorted. We took them off, heated them red-hot to get rid of the distortion and skimmed them, but it was no good. We solved that one. Mr. Goodman - I always called him Mr. Goodman and he always called me Mr. Povey in those days; the only time he called me Fred was when he was very angry - had a brainwave. We found a big lathe and mounted the complete wheel between centres and skimmed the drum in position. We had trouble with the oiling, too.

"The scavenge pump was too efficient for the delivery and gave too much of a dry sump."

I gather that with the inadequate head races and the short and sharp action side-spring Druid forks, the handling was not all that good, either.

But in the race both Povey and Gus Kuhn were eliminated by valve gear trouble. Povey broke a rocker, Kuhn a valve spring collar.

I've no doubt the wiseacres nodded sagely over their beer and said that they knew all the time there was no future in new-fangled ideas like overhead camshafts. But someone took notice for after the race Povey says his mount disappeared mysteriously and was not seen again - not for

Unmistakably Velocette but differing in many respects from production models, this is the prototype overhead-camshaft engine of 1924. Lubrication was constant-loss with pump on end of camshaft. Vertical drive shaft was splined at each end

three years, anyway, when it turned up in Germany.

He had a job to convince sceptics that he had not smuggled it away to a secret power.

Back in Birmingham, the inquest was held. It was pretty obvious from the blued cams and roller followers that the cam gear was having a hard time, and the obvious solution was to reduce the reciprocating weight and the rubbing speed.

Skids replaced rollers and a smaller cam was substituted. The handling was improved by fitting Webb girder forks, then unrivalled for steering and suspension, and modifications to the frame geometry.

"Mr. Goodman and I spent days playing with steering head angles and wheelbase before we settled on 62½ degrees and 52½ inches. The difficulty with the engines was that, although we could run them on the bench until they broke, we could not always find out *why* they broke. We were not getting anywhere much until Mr. Goodman went off to London and came back with a stroboscope. I don't think anyone in Birmingham had seen one before. We soon found out what was happening. With it we could see what happened to the rockers. They straightened out, even bent

1928 KTT — the first three-fifty to cover 100 miles in the hour — seen here in short-distance trim. Note the ribbed crankcase, later standard on production KTT models

sideways. Mr. Goodman designed triangulated ones with the base of the triangle uppermost. We watched the followers not following the cam and he designed a new cam."

(Pushrod racing engines of the period gained much of their top-end performance from the fact that the valves floated - did not follow the cam all the time. Used in an ohc engine, cams of pushrod form either gave poor performance or over stressed the valve gear.)

"The original cam was all right up to 5,000 and we used

Gus Kuhn with the first overhead-camshaft Velocette to be ridden in the TT races

them up in touring models. We saw some fantastic sights with the stroboscope. Saw crankcase walls bending in and out — that's why we had broken crankpins, and cylinder barrels stretching."

I am sure it is safe to say that the intelligent use of this stroboscope (an oscilloscope it was called then) and correct interpretation of the phenomena it revealed, was the most important factor in bringing the ohc Velocette design to maturity.

But the ohc model itself was not the only problem for the Goodmans and their loyal helpers. Once it was apparent that the engine had a future — more future than any two-stroke at a time when the two-stroke was going out of fashion and demoted to the utility role — it was also apparent that the future of Veloce Ltd. depended on its success. Victory in the Island — or at least a *tour de force* — would result in sales demand, and to meet such an anticipated demand they would need production resources beyond the scope of Victoria Street. The very place was available if they could raise the cash to take it over. It was the substantial factory in York Road, Hall Green, then occupied by Humphries and Dawes, manufacturers of the OK. This partnership was about to split up, Humphries to make OK Supreme motorcycles, Dawes to make pedal cycles.

Raising finance was the main preoccupation of the Goodmans at this time and it brought in Harold Willis, whose father was a wealthy master butcher in Birmingham who put up some money. As it turned out, the contribution of Harold Willis to the Veloce fortunes in the next 10 years was far more than could be assessed in terms of money.

Machine tools had to be bought, too, for then, as before and as always, Veloce made more of their machines themselves than most makers.

Early cammies

ALTHOUGH it has been explained that the prototype ohc engine differed considerably from the production models and the unsuccessful TT machines, anyone who has the fortune to examine a 1925 catalogue, and has read the previous chapter, may suffer from some pardonable confusion. There for all to see in the first published details of the ohc range is the model K with the total loss, splined-drive shaft, flat upper bevel cover; and anyone could be forgiven for thinking that I have misled them. The fact is that this was a typical example of catalogue rigging. Publicity and catalogue pictures (the 1924 Show editions of "the books" carried this same picture) were required before a modified model was available, so they did the best they could with one of the prototype machines. The catalogue picture shows an H-type Webb front brake and a box-shaped affair behind the saddle tube which looks more like a tool box than the oil tank one would expect from the accompanying specification, which clearly states "dry sump" lubrication with an oil tank capacity of half a gallon.

Fortunately one of the earliest production model K is in existence — more than that — in *concours* condition in the National Motorcycle Museum in Birmingham. From this machine and later photographs of actual production machines, including one with H-type cast-alloy rubber-covered footboards, we can see what the first production machine — and the 1926 TT models — were really like.

Apart from the engine which, now finalized, remained virtually unchanged until 1935, we find the half-

Believed to be the earliest K model still in existence. The National Motorcycle Museum's beautifully restored 1925 machine

To Fred Povey, man of all trades, it meant going all over the country bargaining for machine tools between testing, riding in competitions and running errands for Mr. John, now reverently referred to as "The Old Man". A forthright, hard-headed business man as Povey remembers well. "When I bought some machinery he said I ought to have beaten them down over the price."

But finally the family had the money and in 1927 they were able to move in, although Dawes stayed on for a time in a part of the building.

They just got into York Road in time, for a sensational success in the Junior TT in 1926 had put the Velocette in the top bracket of British machines — of machines throughout the world — and it was to stay there for many a year.

We have seen how the teething troubles had been ironed out, and by the time the new models reached the Island one or two other improvements had been made. The clutch had been improved to cope with the increased power. Povey recalled that his machine produced 19½ bhp on a 7:1 cr and would do between 86 and 88 on the road. It now had three Ferodo inserted plates instead of the original single corked plate, and the old scissors action quick-thread worm operation for the clutch, descended from the two-strokes, was replaced with the self-aligning ball thrust operated by a tilting lever and a push-rod in the gearbox, as used to the end of production.

The oil feed pipe was now ½in in diameter. The non-saddle type "wedge" tank had a big, quick-action filler and was mounted on big rubber pommels on substantial lugs and the lower tank rail. It was stressed that the engine was standard apart from the cam and the piston.

The team was Alec Bennett, Gus Kuhn (who soon abandoned road racing for a successful speedway career) and Povey. The appearance of Alec Bennett on the Ve-

gallon oil tank on the seat-tube shaped rather like a very fat first war water bottle, a rather unlovely welded steel silencer (painted black) and an attractive front brake of new design. This front brake, a Percy Goodman design, set a new standard both in efficiency and appearance. The cast-alloy brake plate, supporting generous (for the period) alloy shoes, had a substantial built-up boss which made a first-rate bearing housing for a long cam spindle and was flat on its leading edge to butt firmly against the back leg of the girder forks. The anchorage, therefore, was immovable without extra bolts or struts. A small detachable alloy block in front of the fork blade prevented backward movement.

Finally the lower edge of the brake plate was flared out and extended clear of the brake drum to lead off any water which might run down the plate. The whole design showed that Percy had a keen appreciation of the importance of a good front brake (not all designers shared his conviction at that time and many riders thought front brakes dangerous) and an engineer's understanding of the technical requirements. So good was this brake that it was adequate even for the racing models up to 1931, when the drum was strengthened and widened. One or two minor details are not so apparent. The gearbox and clutch with single cork plate are virtually the same as on the two-stroke model H, clutch operation being by quick-thread worm behind the clutch. There is no internal indexing of the gear-selector mechanism, reliance being on the tank-side gate to locate and hold the gears.

The early catalogue pictures show a shapely cast-alloy silencer with a very short tail pipe sliced off at 45 degrees, virtually a small Brooklands style silencer without the characteristic (and compulsory) fishtail. I feel sure I have seen photographs of early K models in trials with this silencer, and one of the later publicity photographs shows it, though a companion picture of another machine (a different machine because the saddle is of another make and the kick-starter crank lies at a different angle) shows the unlovely steel box.

The National Motorcycle Museums' machine has an authentic steel box and you can judge for yourself if I have been too unkind. However this machine, with the delightful registration number TT 5669, was registered at Exeter in September 1925 and bears the serial number K 23. Presumably it was made in July 1925 because Bob Burgess acquired number K 33 in August. Even supposing Veloce Ltd. started numbering at number one — very unusual in the case of production machines — this must be a very early example. I would not surprised if the numbering started at 20. The steel box persisted for the whole of the vintage period but in a more shapely form. It was deepened until it resembled another army water bottle tipped by a stub tail pipe like the cast-alloy one and mounted farther rearward so that the tail pipe hid coyly behind the rear fork end and produced that deep thump of an exhaust note by which you can still identify a Velocette.

locette was a tremendous compliment to Veloce Ltd. He had approached them in the first place, not the other way round, so confident was he that the Velo was a potential winner. The only difficulty was that Bennett was a professional, at the top of the tree and in a position to demand a considerable sum for his services. Far more than Veloce Ltd., already straitened by their factory commitments, could entertain. Agreement was eventually reached by negotiations aided by the fact that both sides wanted a deal — an agreement which, I gather, was kept from Mr. Goodman, senior, as long as possible. I was also told that he was so huffed when he did find out what the sum involved was that he didn't go near the factory for weeks!

The agreement was that Alec would get what he wanted if he won but next to nothing if he didn't. He must have felt he was backing a cert.

His race tactics were a careful blend of confidence and caution. Wal Handley on the Rex Acme Blackburne, now at the peak of its development and speed, was the real danger.

You could bet that Handley, who had won the race the year before and held the lap record, would in his usual manner set off like a rocket to demoralize the opposition. Perhaps the spidery Blackburne, representing the old school of pushrod engines which owed more to development and tuning than empiric design, would blow up. Until it did Handley in his inspired moments was the most tigerish rider the Island had seen. The very sophistication of the Velocette design with its dry-sump lubrication and full pressure feed to the vertical shaft and cam box imposed a limitation on flying starts. There was a lot of oil to be warmed up before it was safe to use full revs and so much drag that the engines did not give of their best for at least a lap.

Bennett with the restraint of caution and confidence took it very easy on the first lap — never, he said afterwards, exceeding three-quarter throttle — and then filled up. Handley, as was expected, was away in the lead. Bennett was content with third place. On the third lap, the oil now nicely warm, he turned up the tap, skittled Handley's record of 65.2 mph with a lap at nearly 69 mph and took the lead.

Eric Thompson's 'other' Velo. The former Vintage Club secretary captured in full flight at the Royston Hill Climb on his KTT

Congratulations for winner Alec Bennett from Percy Goodman after the 1928 Isle of Man Junior TT

He filled up on the fifth lap to be told that he had it in the bag, but instead of rolling it back and taking it easy he turned it all on to show the world what the Velo could do. Sheer confidence, this must have been, yet it nearly brought disaster for he came off at the Nook, cutting his chin but not doing any damage to the machine.

Despite this delay he won by 10 minutes 25 seconds, at 66.70 mph Jimmy Simpson was second on the AJS at 63.9 mph, just pipping Wal Handley. Gus Kuhn was fifth, Fred Povey was ninth.

They carried Percy Goodman — a shy but very happy man — shoulder high round the paddock. For they knew that this was a triumph for a designer, not just the success of a skilful jockey and a tuner. It was said that all the Velos were as alike as peas, that Bennett was not given the best one, and everyone believed that he would have won on any of them. It was, we can see now, a prophetic victory, the beginning of the end of the era of the proprietary engine in an assembled machine.

Inevitably when one searches for words to describe a historic happening like this, one finds that the great prophet and scholar, Ixion, has already used them. Thus it was when I sought to analyse the cause and effect of this Velocette history. "This machine is well bred by brains out of honesty" wrote the great man with an economy of words which makes further eulogy unnecessary. "Its glory will not be dimmed easily by inevitable runs of bad luck," he added prophetically. A prophecy which was to come true much sooner than he could have imagined!

To qualify his statement and his prophecy, Ixion used one of his familiar devices. He quoted a "rival manufacturer" as saying: "Veloce have always disdained the cheap and nasty. They have steadfastly refused to buy any but the best stuff or to tolerate anything but the finest work. They would rather let other firms undersell them by £15 or so than descend to the second rate, and I think they are going to get their reward at last." Now a rival manufacturer may have said something like that, but the words are pure Ixion. He added another prophecy which was to come true, and is still true, as the continued success of the Velocette Owners' Club testifies. The Velocette would, he said, start a cult like the "Scott... each year it would breed an increasing number of men who instead of changing makes each year would ride Velocettes and nothing else". So of course it was. Dedicated riders grew up and dedicated dealers, too, many of them rider-agents, and on this firm foundation was built the final success of Veloce Ltd.

There was, in my opinion, only one point that Ixion missed in his appraisal of the vintage Velocette. *The* reason behind the reasons he mentioned. That was, I submit, the fact that everyone of real influence in the firm was a practical motorcyclist. Percy and Eugene regularly rode their own products — Eugene often riding in trials. By all accounts they were critical testers.

Bob Burgess recalled:

"E.F.G. (Eugene Goodman) rode in many reliability trials but I never knew P.J.G. (Percy Goodman) ride in competition. He used to go to and fro from home on his products from time to time. But E.F.G. used solo and sidecars daily at one time. Two riders more brutal on the machinery I have never known: they certainly found out any weak spots. In this connection I once asked E.F.G. if I might borrow a new model that the firm had on test. He replied: 'No, you are no good — you never break anything.'

"This shows," Burgess continued, "that the brothers were always willing to sample their own medicine and were both tolerant of intelligent criticism of the products from members of the staff, particularly P.J.G., who was perhaps the most critical of his own productions. He was a perfectionist and constantly strove after improving the machines and the efficiency of the factory. George Denley (later sales director) also rode in competitions, including the TT."

And right down the line the men who made Velocettes in those days were enthusiasts and riders. Fred Povey, Harold Willis, Charles Udall and Bob Burgess were all ordinary motorcycle club lads who found their way to Veloce like pilgrims to Mecca and stayed on throughout the formative years, gradually rising to executive positions. That was why all Velocette machines had that indefinable quality which can only be described as "a rider's machine".

When the Velocette team went over to the Island in 1927, it was with quiet confidence. A few detail improvements indicated that they had learned from past post mortems and a few more horses had been found. Lubrication had received most attention. The feed pipe had been increased in diameter to prevent starvation when starting from cold, and so that more oil could be fed to the cam box

The first 'amateur' Velo rider in the Island, W.S. Empsall with his very standard-looking K in the 1926 Amateur TT. He won the 350 cc class at 52 mph, finishing eighth overall

to cool the highly-stressed cams an extra pump was fitted on the end of the camshaft, to remove the surplus. A new top bevel casing with more clearance reduced oil drag. A new steering-head lug took much larger races to cope with the reaction of the powerful front brake. A trifling modification was made to the split sleeve which, screwing into the rocker arm, locked the tappet adjustment. Originally it had been split all the way, now it was only split part way to render slacking off for adjustment more easy.

The team was Bennett, Longman and Willis. Willis had not been long with the firm but already showed a streak of obstinacy for which he became well known. His motto was "Believe nothing you are told and only half of what you see", a principle which was followed by himself and Charles Udall throughout the later racing years when wild stories and duff gen. were regularly distributed by rival camps. Willis had his tappet sleeves slotted *all the way.*

Practice was uneventful, with either Bennett or Willis topping the lists most of the time except when Wal Handley, having broken his Rex Acme Blackburne, borrowed a Velo (he was a close friend of Willis) and put up the fastest lap of all. That was Wal all over.

As expected, Handley went off like rocket in the race, with an opening lap in 33m 11s. Freddie Dixon on an HRD was second, nearly a minute behind, followed closely by Charlie Dodson, Sunbeam, and Alec Bennett, cautious as ever. Longman was eighth, Willis 11th. Lap two and Handley got down to 32m 44s, breaking Bennett's last year's record, drawing further away from Dixon. Dodson had a slow puncture and Bennett nipped in to third place.

Joan Makin checks the route before setting forth on her 1925 Model B

The third lap was the same story, with Bennett slowly catching Dixon, and the fourth lap a repeat except that Bennett seemed to be slowing. On the fifth lap Handley threw it down the road at Quarter Bridge but got going again, still in the lead: Dixon was still going strong but Bennett was in trouble. The modified tappet adjuster had slacked off, the clearance gradually increasing and slowing him until the hammering broke the rocker. Longman had the same experience — first a loss of revs and then a full stop with a broken valve collar. Jimmy Simpson came into third place on the cammy AJS and Willis in sixth place was left to carry the Velocette banner. He had his problems too for at his first pit stop his attendant strained the hinge of the oil filler-cap and it would not close.

Desperately Willis ripped the cap off complete and stuffed a handkerchief into the hole. Despite the worry of losing his oil, Willis responded nobly to the flat-out signals.

The last lap was packed with drama. Handley, leading by two minutes, went out with a busted piston, Simpson had his rear chain come off. Willis grabbed second place in the last-lap confusion, only a matter of seconds ahead of Simpson and Reynard on a Royal Enfield. Dixon won at 67.19 mph, Willis averaged 64.78 mph, not a particularly high speed but his role had been to support Longman and Bennett and he had done it well. His hunch over the tappet adjuster had paid off.

The Velocette tappet adjuster is a clever piece of work and was copied by Nortons when Carroll redesigned the camshaft Norton. The tappet foot has a tapered stem and a final threaded portion for a nut. This is assembled in a threaded sleeve which screws into the rocker. The sleeve is tapered internally to match the tappet and is saw cut for the whole length on one side with three other saw cuts to the end of the thread to make up a cross section... literally. Adjustment is made by screwing the threaded sleeve in and out of the rocker, and the tappet foot is then drawn up into the taper by its nut. As it is drawn up the taper expands the sleeve locking it in the rocker. The tappet foot cannot turn because it has a flat engaging with a projection on the rocker. To release the assembly the nut is slacked off and given a sharp tap which releases, or should release, the sleeve. The main snag is that after the tappet has been hammered a few thousand times by the rocker it can be driven further through the sleeve, the adjustment goes haywire and the nut

Alec Bennett's TT rides for Velocette

"PERCY and Eugene Goodman were, as you must know, brilliant engineers. On examining the 1926 engine on the bench and watching the power curve and the test that it went through, I was convinced that this engine was a winner, coupled of course with a good frame design.

"I came up with the gate gear change on Sunbeams, and after riding for Norton the foot change (non-positive) was naturally a great improvement and a saver of time, especially on the Isle of Man circuit. When Velocette produced their foot change (positive stop) in 1927, which was an improvement on the Sturmey Archer, this again saved seconds per lap.

"The modification to the tappet adjusters on the 1927 machine, a solid taper instead of a split cone which was used in the 1926 race, caused the adjustment to work loose in the 1927 Junior Race and, I can safely say, lost me the race."

"The reason for my front brake being worn at the end of the race was probably due to going deeper into the various corners on the circuit and braking hard."

"I preferred using a trigger lever instead of a twistgrip as this particular lever was geared very high (a large cable drum) and could be used with just the index finger without movement of the wrist or hand.

"You may be interested to know that my mechanic for the 1926 Junior TT was Mr. Percy Goodman himself. He did complain to me at the end of the first week that he didn't have enough work to do on my machine. As a matter of fact he didn't put a spanner on it during the practice period but the day before handing the machine in he removed the cylinder head to check the valves and bore, etc., and then reassembled it.

"It was one of the first dry-sump designs (the 1926 machine) and the feed from the oil tank to the pump was by ¼in copper pipe, and with the heavy Castrol R that we used it took just a lap to warm up the oil. This meant that on the first lap you went along on half throttle, which put one well down the field on that lap, although I never did push my machines in any way as I considered that warming up the metal parts gradually gave one a perfect motor which came in very useful towards the end of the race."

slacks off. Rechecking after running and self-locking nuts are the remedy, after which the tappets will run long distances without attention.

Bob Burgess who had the good fortune to follow the activities of Velocettes in the Island from 1925, first as a private owner and later as a Velocette employee, summed up the 1927 tragedy thus: "A Junior race almost certainly lost by the failure of such a small item (cost then 1s 3d) is surely a similar case of that of the traditional loss of a kingdom for want of a horse-shoe nail". Willis, he added, showed him his second-place cheque with a characteristically Willis-dry quip, "This'll deal the overdraft a nasty crack".

Burgess had another recollection of those early days which must have warmed the cockles of his heart. The year before the works team were using a modified hand change. Selector indexing was provided in the gearbox and the tank-side gate deleted in favour of a spring-loaded stop on the lever pivot to reduce the chance of missing second gear. The idea was to speed up gear-changing but Bennett did not like it and asked for the normal gate change. The only K model in the Island was Burgess's holiday mount and he volunteered to loan its gate change, which was hurriedly fitted to Bennett's machine. Few enthusiasts can look back and say, "I helped a TT star to victory".

For 1928 engine modifications were minimal. The compression ratio was up to 7.25 to 1, the rockers were stiffer. The upper bevel cover was more bulbous. Rear tyre size was increased from 2.75in to 3in. The secret weapon which was expected to improve lap times without resort to more engine tuning was the Willis-designed positive-stop foot-change.

Foot-changes were not new, they were as old as three-speed gearboxes, but previously they had been a mixed blessing because it was all too easy to miss the middle gear altogether when booting the pedal. Only Nortons had got round that problem by craftily arranging the internal ratios so that the bottom gear train was a much higher ratio than second. Which meant that the middle ratio, being at the end of the pedal travel, could not be missed. It involved going through bottom (the old middle ratio) on the way to the new middle but this was no problem with the Sturmey Archer dogged engagement. But it was only a dodge and served only to emphasize the need for a positive mechanism which would enable the insensitive boot to make accurate changes. This was the problem that Willis solved by, it is believed, adopting a ratchet mechanism used on a German made slotting machine in the Hall Green factory.

Quite apart from positive selection, the great advantage of the positive-stop change, the lasting memorial to the Willis inventive brain, was that the pedal always returned to the neutral position, which meant the riding position could be compactly tailored to the rider's stature without allowance for "knees up, Mother Brown" antics. The subsequent production Velocette positive-stop unit, standard equipment on the first KTT model in 1929 but available as an extra for other models and later sold to the Scott company and used on 1960s Birmingham Scotts, is well known and differs only in detail from the original Willis unit. Released as a surprise in the Island, the original one had a spring-loaded face cam to centralize the pedal, a system which Sturmey Archer adopted when they evolved

First experimental Velocette four-speed gearbox was built in 1929 and used on works machines. Willis positive-stop unit operates plunger of rack and pinion to rotate snail groove selector drum

The experimental four-speed gear cluster

their own ps foot-change the following year. There was also a spring-loaded "dash pot" in the rod linking the foot mechanism to the gear selector so that the pedal could be fully depressed without waiting for the gears or dogs to mate. The gearbox, it should be noted, was still basically the old two-stroke unit improved by the substitution of ball races for pb bushes on the layshaft and with the gears strengthened by a change in tooth form.

There was not much pre-race publicity given to the positive-stop foot-change... the limelight was taken by the spring frame Velocette which appeared in practice. This had a Bentley and Draper-designed pivoted rear fork grafted on to a K frame front section, a design virtually identical to that used by George Brough for several years and later imitated to some extent by Matchless and New Imperial. Not unlike Philip Vincent's spring frame, too. The peculiarity of the B and D design was that the seat stays did not contribute much to the rigidity of the rear fork assembly, being pivoted to the rear fork ends with friction dampers interposed and connected to the main frame by short pivoted links (girder fork links) under the saddle where they reacted on the short compression springs which provided the suspension. The rear pivoted fork pivoted on small plain bearings on a through spindle but the only stiffness of the assembly came from a triangulated sub fork below the pivot with a brazed-in cross tube.

Looking at it now, you'd say it was rather a clumsy arrangement and difficult to justify for its complication. But this was a long time ago and considerations of easy adaptation to existing frames may have influenced the designer. I am sure it was no more than an experiment from Veloce's point of view and not a serious TT contender. For one thing, the congestion of links and springs under the saddle ruled out the normal oil tank and the special triangular tank evolved for the limited space precluded fitting a positive-stop change. The engine, too, was a last year's unit which had seen service on the Continent. Practice demonstrated that there was not enough cornering ground clearance on

"Spring-heeled Jack"

WILLIS dubbed the B and D spring-heel Velo "Spring-Heeled Jack" and this nickname has handed down through the years, like so many Willisisms. I only had a vague idea who or what the original Spring-Heeled Jack was — assuming that it must have been the name of some unusually agile sportsman. In fact the origin is a bit more interesting than that and bordering on the supernatural. Spring-Heeled Jack was the name given, in the absence of anything better, to an apparition which appeared in various parts of the country between the years 1837 and 1904, causing even more consternation than flying saucers might do today. The story began in that earlier year with the finding of a man battered to death on Barnes Common in London after a mysterious series of robberies and assaults.

Because of the assailant's ability to escape from the scene of the crime the newspaper coined the name. The first accurate sighting came from a girl who was attacked at the door of her home by a stranger who first posed as a policeman, saying, "We have caught Spring-Heeled Jack down the lane". The man or creature then threw his cloak aside to reveal a thin body in a tight-fitting oilskin suit with a gleaming metal helmet and spurting white and blue flame from his or its mouth.

Various similar reports came from other parts of the country over the next 30 years until an incredible encounter was reported from Aldershot. The figure, still in the tight-fitting suit with the metal helmet, surprised two sentries guarding a military powder magazine.

One sentry dropped his rifle in fright and Spring-heeled Jack leapt over him. The other sentry opened fire at point blank range with no result except that "Jack" spat blue flame. The other sentry then fired but "Jack", unperturbed, jumped on to the top of the sentry box. Four more soldiers and a sergeant joined in the hunt, firing repeatedly at "Jack" who, spitting flame, danced about the roof tops.

Not a drop of blood or a single footprint was found next day. The last sighting was at Liverpool in 1904 where around one hundred people testified to seeing Jack, dressed as usual and still spitting blue flame, cavorting along a street, leaping as high as 25ft at a time until he made off across the roof tops. The only reasonable explanation of the phenomena was that Jack must have been a circus artist accomplished in acrobatics and flame-swallowing equipped with a bullet-proof vest and spring-loaded shoes. Add together the black, form-fitting suit, the protective helmet, the flame spitting and the agility, and you can see that as usual the Willis coined nickname was apt.

full bump, the low-slung sub fork being further complicated by a central stand pivoted to it.

Rumour had it that the spring frame would not go into production though there was eulogistic press tests of it. Factory comment was that it might be produced when manufacturing capacity permitted. Rumour, as it turned out, was correct. The inside story was that although the question of ground clearance could be overcome, no effective way of carrying a pillion passenger could be seen and that, to Percy Goodman, was enough to rule it out. The firm's policy was still unshakable. They would only pursue racing developments which would improve the breed of their road-going machines. Much later, when racing demanded a spring frame (at the insistence of Stanley Woods), the conflicting requirements of racer and roadster were satisfied to the company's satisfaction by Phil Irving's design of a pivoted fork rear suspension adjustable to the extra load of a pillion passenger — a design used on all successive spring-frame Velocettes.

Practice showed that other makes had found more horses. Simpson (AJS) lapped at 66 mph, Dodson (Sunbeam) at 67.7. Willis then had a real go in good conditions and returned 69.2 mph Handley, incidentally, scratched because his Rex was not fast enough. Dodson set the pace in the race with an opening lap at nearly 69 mph with

The one-off 1928 spring-frame Velocette tried out in TT practice but soon abandoned. The machine was restored in Birmingham in the late 1960s. This close-up shows clearly how the Bentley and Draper patent spring frame design was incorporated into the traditional Hall Green layout

Michael Tomkinson

A YOUNGER generation may remember Michael Tomkinson as the man who built the way-out Laverda-powered hub steering endurance racer "Nessie", such is the power of a press nickname; but Velocette enthusiasts will remember him for his regular entry of a Venom in the prestigious Barcelona 24 hour race. Second place was the best result, victory always eluding the team but not for want of meticulous and imaginative preparation. Long before that he had prepared an innocent looking 1927 vintage KSS into a veritable giant killer in sprints and short circuits, unbeatable in its class or any vintage class and, with regular rider Howard German up, capable of leading a field of modern racers for several laps. Second fastest motorcycle times of the day at Shelsley Walsh hill climb in

Showing something of the concentration on the job in hand which took him to success in the preparation of long-distance race Venoms, the BSA B50, the revolutionary "Nessie" racer and in the motorcycle and car business, the late Michael Tomkinson is pictured on his vintage two-stroke Velocette in the first Vintage Club Banbury Run, in 1948

Arcangeli second on another Sunbeam and Simpson third. Canny Alec Bennett lay sixth on the first lap but then got down to business, set up a new lap record of 69.53 mph and jumped into the lead, Dodson having run out of oil. Willis moved up to fifth but neither newcomer, Brooklands star Freddie Hicks nor old-timer Povey were as yet on the leader board. Bennett then lapped even faster and the strain was too much for the AJS of Simpson (AJS had reverted to push rods and ran into valve spring trouble), so Willis moved up to second place. Bennett broke the lap record for the third time and won easily at 68.65 mph, Willis was second at 67.16 and the third man, Ken Twemlow (DOT-JAP), was miles behind at 63.67 mph

It could be seen that Bennett's record second lap had burned off the Sunbeam, Norton, AJS challenge which faded out after the third lap, yet the finishing Velos (Hicks was fifth but Povey retired with — you've guessed — a broken tappet adjuster, overtightened, it was said) finished in perfect mechanical condition. The only adjustment necessary was to Bennett's front brake, which, not surprisingly, was well worn down. Bennett said afterwards that he had opted for a higher top gear than the others and, running into a high wind, had to get down to it more than usual. His fifth lap record of 70.28 mph broke the 70 mph barrier for the first time by a three-fifty.

There was also a wonderful demonstration of the stamina of the cammy Velo when Hicks won the 350 cc class of the 200-mile sidecar race at Brooklands at 70.84 mph, a speed which would have won him third place in the 600 cc class and second place in the 1,000 cc class. Hicks was responsible for much of the track development, aided in the test shop ('din house' in Willis phraseology) by Willis. A successful season in 1928 culminated in Willis taking the hour record at 100.39 mph, the first man to top the century for an hour on a three-fifty. At Montlhery, without the restriction of a silencer, world records were taken up to 12 hours. The racing history of the company is well summarized in L.R. Higgin's invaluable book, *Britain's Racing Motor Cycles* (Higgins having been a successful Velo private runner). He gave the settings for this 100 mph engine. Compression ratio was 10.5:1 and valve timing: inlet opening 38 degrees before tdc, closing 45 degrees after, exhaust opening 60 degrees before bdc, closing 45 degrees after. About this time there was one day a loud

---

1957 was a remarkable achievement for tuner Tomkinson, rider German and the 1927 cammy Velo. FTD went to a 499 cc Featherbed Norton, a mere 1.5 seconds faster. Tomkinson was a founder member of the Vintage MCC and approached the early speed events with the same careful logic that characterised his life.

I remember telling him, back in 1947, that a KSS Velo (a Mk. 1) for the 350 class and an Ulster Rudge for the 500 class were the best machines for our early speed events. He knew about cammy Velos because he had ridden mine in the early thirties and had seen what Rudges could do on dirt and grass. It was not long before he found both and rebuilt them. He rode the Velo nearly a 100 miles to show it to me and run it in. With the compression in double figures for alcohol the thin base barrel let go but Vintage Club regulations would not permit use of the thick barrel introduced for the KTT in 1932. A brilliant engineer, he made an exact copy of the vintage barrel... but in steel.

When he gave up entering riders in vintage events he sold his collection of vintage Velos to Ivan Rhodes who continued the campaign.

1927 ohc Model K, production version of the mount which gave Alec Bennett a runaway victory in the 1926 Junior TT at 66.7 mph. In this form it was used on the road by Michael Tomkinson before being developed into a "giant killer" which "blew off" post-war racers, ridden by Howard German. Now owned by Ivan Rhodes, it dominated vintage races.

bang, and a hole appeared in the tin roof of the test shop. A long-suffering cast-iron cylinder barrel, provoked beyond endurance by higher and higher compression ratios for dope fuel, had let go at the base.

Barrels with a thicker base were made up and used for these Brooklands motors although it was not until after 1931 that thick-base barrels were put into series production for KTT owners, and then only because private owners were experimenting with ultra-high dope ratios.

The TT models were still "same as you can buy", the 1928 winner being, apart from the foot-change and "careful assembly", the KSS model, a replica of the unfortunate 1927 TT mounts. The KSS (K Super Sports) was sold with a guaranteed maximum of 80 mph on a compression ratio of 7 to 1 for petrol benzole fuel, though I suspect you needed a straight-through exhaust pipe to do it. A low ratio of 6 to 1 for straight petrol or 8½ to 1 for Discol was optional and, normally, close gear ratios were supplied. It was, in fact, a real TT replica though not described as such., and private owners bought and raced them with great success. List price was £75 — £10 pounds more than for the cooking K and KS (K Sports models). A cheaper version, the KE (K Economy) sold at £58. Cost was saved here by the use of a pressed-steel cover for the primary chain in place of the elaborate cast-alloy oil bath on the more expensive models. Older designs of cam, rockers, cam box and cycle parts were "used up" on the KE model.

Confusing to the collector is the fact that symbols on the engine crankcase did not always tally with the model designation. A case in point I recall from my teenage days...

A more prosperous friend had a Velo which I now know to have been a 1928 KSS — 2¼-gallon petrol tank with quick filler, 21in wheels fore and aft, and narrow guards... a very desirable property. The engine was stamped KCR, which we all interpreted as being very racy. I often wondered about it and quizzed Bob Burgess, who as service manager was in the best position of all to sort out the profusion of symbols. He explained that the C meant that the engine had the latest type of cam (the 80 mph one) and the R signified the latest type of strengthened rocker. A K-type engine with the then last word in goodies.

While on the subject of nomenclature I will move on to the next year, 1929, when there was a tidying up of the roadster range and the introduction of a new letter. Models K and KE were both replaced by model KN which had the frame and forks of the 1928 KSS, and the sports models KS

Harold Willis (left) and Alec Bennett, 1928 Junior TT

and KES were both replaced by the sports KNS. The N indicated a machine with all the NEW features, some previously indicated by symbols like CR. There was still a KSS model, virtually unchanged, but the role of leader of the clan had been taken over by the aptly designated KTT.

The KTT was offered to the public as "The fastest 350 in the world" on the strength of 60 world records obtained in 1928, which included the hour, 100 miles and 1,000 miles, and was described simply as an exact replica of the machine which gained first, second and fifth in the Junior TT and broke 60 world records all in the space of a few months.

Split second before Ivan Rhodes guns the ex-Tomkinson KSS off the line at Packington in 1958 to make another FTD

The KTT: for beauty of line it has no equal
An honest attempt to give the private owner a pukka racer

Freddie Frith pushes off in his first major road race — and the first Manx Grand Prix — on his over-the-counter KTT in 1930. He finished third, at 60.34 mph. Velocettes filled the first eight places. On his three-fifty Frith lay third in the Senior Race at 68.46 mph until it blew up. Note two pairs of spare goggles

The 1929 production KTT was not an "exact replica" of the 1928 TT winner; it was slightly better because it had an improved Webb fork with a bracing tube each side between the lower forkbridge and the fork end.

This strutted fork, exclusive to vintage and early post-vintage KTT models though not used on the 1928 TT mounts which had normal racing Webbs, was a finishing touch to a machine which for fitness of purpose and beauty of line has, in my humble opinion, no equal in motorcycling history. There have been more imposing machines, more exotic machines, more flamboyant racing machines but never one evolved with such honesty of purpose. Nothing was added for effect or for sales appeal and nothing was omitted for price cutting. It was an honest attempt to give the private owner a pukka racer, a replica of the last Junior winner at a price he could afford (£80 against a KNS at £62... good examples of both are worth many times those prices today !).

There was more to the decision to market the KTT than simple money spinning. Road racing was moving into a critical period. The chill wind of depression was blowing through the factories and manufacturers were dropping out of racing. The ones remaining were signing professional and semi-professional riders and there was no room for the amateur. If a rider could not afford to have a go on his own,

A 1930 KTT. Unquestionably the most successful production racer of all time... and still the most desirable vintage racer

no maker would any longer sell him a replica of a competitive works racer. Not until the KTT Velo...

By producing these machines in considerable numbers, Velocette gave road racing a real shot in the arm, the extent of which we shall see later.

But back to 1929 for the Junior TT which put the final seal of success on the machine designated the KTT.

The origin of the strutted fork is interesting. No one at the works could remember how it came about but Bertie Goodman suggested I contact Webbs, who have long gone out of the motorcycle world to find new fame as manufacturers of rather superior grass-mowing machines. And, surprising in these days of take overs, I was still able to talk to one of the original Mr. Webbs. He told me that back in the early 20s when there was a Sidecar TT there was a demand for a braced front fork to cope with the side thrust. Other manufacturers and sidecar racers added rather unsightly bracing members on the outside of the forks, usually running from fork end to top spindle pivot and tensioned by a strut at bottom spindle height. A bit of an eyesore, I think, and so thought Webb's designer (regrettably his name escapes me) who evolved bracing which was neat and inconspicuous... the tubular strut between fork end and lower fork bridge; an inside strut instead of an outside one. The demand soon evaporated with the death of the Sidecar TT, though it was briefly revived by sidecar speedway racing, and this was probably when the design came to the notice of Velocettes who were mildly interested in speedway, enough to produce a short-lived speedway model with a special frame and a bored to 411 cc dope motor. It would have been difficult for them to ignore speedway for the Hall Green track was but a stone's throw away. They chose the new Webb speedway fork for their "dracer", a horrible abbreviation popular at the time, a telescopic design based on the Harley Peashooter fork, but took a fancy to the strutted girder for the KTT. A very strong fork it is and not easily bent. The snag, as Webbs found, is that it is extremely difficult to straighten when it is bent.

The works team for 1929 was Bennett, Willis and

Sole surviving dirt track model

"Family" group with an international flavour in the Isle of Man in 1930. Percy Goodman and sister Ethel are seated front right; Harold Willis is standing directly behind the rear wheel of the KTT. Other riders represent nearly all the countries that sold Velocettes

Freddie Hicks who had won his way in by reason of his great work at Brooklands. He was a dashing rider and a meticulous mechanic who always liked to prepare his own machine. Private runners included Don Hall, who won the 200-mile South African TT, Sid Crabtree and Tommy Simister. Of these all but Hall and Bennett entered for the Senior as well, for good measure. In addition to these "star" men there was an amateur clubman, W.S. Braidwood, and J.W. Shaw, making his last appearance. A total of eight KTT models of which seven were to finish without any mechanical trouble and one was to crash... and that the last one you would expect. Hicks showed his form in practice by putting up the best lap of the team in 33m 30s but the redesigned Carroll-type ohc Nortons proved very fast and Jimmy Simpson took the first week practice "record" with 33m 9s. Handley, on an ohc AJS was dangerous too, but tipster Ixion of *The Motor Cycle* suggested that a lot of the practice times had been camouflaged and preferred Dodson to win on a Sunbeam, with a Velo in the first three. After this prognostication had been written Simpson put in a flyer at 32m 23s and Alec Bennett, who had spent the first week practising on his Sunbeam, whipped his Velo round in 32m 59s to show that he had not forgotten how to ride one.

The race did not run to form for although Simpson led on the first lap at record speed his front brake seized on and a stop to free it put him off the leader board. His ruthless attempt to win back his place soon blew up the motor. Stanley Woods, second on the first lap, did not last long, either, and Hicks, third at first, soon found himself in the lead with Bennett second and Handley third.

Willis lay fourth but fell on lap three and damaged his bike. Handley broke the lap record and got by Bennett in his pursuit of Hicks, gaining six seconds on one lap. Hicks got the message and turned more on, putting in a record lap at 70.95 mph and getting 41 seconds in hand, which was too much even for Handley to make up, especially as he was ahead on the road.

Hicks very nearly overdid it on the last lap, though. Skidding on wet tar at Glen Helen, he grazed along a wall and bent his footrest but did not come off and came home a popular first-time winner at 69.71 mph, winning by 82 seconds. Bennett was third, a minute or so behind. Dodson was a poor fourth (the Sunbeam times had perhaps not been camouflaged), Simister was fifth, Don Hall sixth, Crabtree seventh, J.W. Shaw 10th and Braidwood 11th (he had fallen off, torn his leathers and bent his machine but gamely carried on).

In the Senior the Velo men were praying for a downpour to even their chances. They only had showers but nevertheless really got among the big bikes. Hicks was on the leaderboard by the third lap and while the Rudges and Sunbeams fought it out, with final victory to Charlie

Velocette agent for Japan, Kenzo Tada was invited to ride in the TT in 1930. His only Island ride netted him 15th place in the Junior, after many practice mishaps and tumbles that earned him the nickname 'the India rubber man' for always climbing back in the saddle after a fall

Dodson backed by Alec Bennett, Hicks worked up through the field to finish sixth. Crabtree was 12th and Ernie Thomas, in his first Island ride, 14th. Fred Craner, soon to give us Donington Park circuit and become one of the best loved (and feared) characters in the sport, provided the most spectacular tumble. His Velo cast him off at Ballaugh (he used to say afterwards that it seized) and he performed three somersaults, the last of which landed him conveniently at the door of the Raven Hotel, into which he was promptly carried.

Not a bad show all round for the new "same as you can buy" racer.

The only noticeable differences from the Show models was the substitution of a single leather-lidded tool box on the off-side chain stay for the twin tool bags high on the rear mudguard stays. It was claimed that all Hicks did to his Junior mount in preparation for the Senior was to lift the head and grind the valves. No replacements, not even the tyres, were necessary and all the placed Velos stripped to perfection. As usual, Bennett's front brake was worn down more than the rest. Incidentally, he still preferred his pet trigger lever to a twistgrip for throttle control. It was like an air lever but set so that it shut when released.

The year 1930 was the first year of the "foreign invasion" on Velocette machines. Willis was the only representative of the factory and he was trying out an experimental model, but the lists were full of private owners on KTT models, many of their names unpronounceable, most of them somewhat off-put by the blind corners and stone walls. The Willis experimental model was interesting. For one thing, it had a four-speed gearbox for the first time, and for another it was a long-stroke 68 x 96 mm, instead of the traditional 74 x 81 mm. The idea was to get a smaller, more compact combustion chamber but, in fact, there was no appreciable gain in performance. The dimensions did come in handy later on, but that is another part of the story.

The gearbox marks the beginning of a new chapter in the gearbox story. It had become obvious by 1929 that a better gearbox was required, and that it would have to have four speeds. Till then the strengthened "two-stroke" box, but with the addition of internal indexing, had served well but was getting near the limit of reliability and suffered from the disadvantage of a crash second gear instead of dog engagement. Not ideal with foot change. Willis therefore evolved an experimental batch of four-speeders, of which a few were made for works machines. Novel for Velocette was the revolving drum selector with snail grooves for moving twin selectors — very similar to later Burman boxes. The drum was rotated by a rack and pinion, the rack emerging as a plunger on top of the box, exactly like an early Sunbeam box. The plunger was operated by a slightly modified Willis-type positive-stop mechanism. There was no provision for a kick-starter. This gearbox seems to have worked quite satisfactorily — in fact, the one remaining example was later frequently used by Grahame Rhodes in vintage races today.

Eugene Goodman next had a go at designing a four-speed box. Being more of a road and trials man, he had different requirements in mind. For one thing, he wanted a higher kick-start ratio than the old design permitted so that the engine could be spun. And while he was about it he wanted to get rid of the old roller-type kick-start engagement mechanism which had caused some dissatisfaction... could be he had experienced that awful jar up the marrow when, having thrown all your weight on the kick-starter crank for that long swinging kick beloved of writers of the "Torrens" era, the roller slips and the crank goes down with no resistance whatsoever. Anyway, Eugene plumped for the more conventional ratchet wheel and pinion, like Burman and others. While he was about it he also thought he might lay the Velocette ghost of the mysterious clutch operation (the uninitiated can never understand what goes on after the clutch cable has disappeared into the top of the gearbox on the offside. It was a long time before I did). To help, Eugene had an external lever to operate the normal clutch mechanism. Heavier shafts and gears were obvious requirements.

The result was a rather bulky gearbox, probably no bigger than some contemporaries, but because the narrow chain line was not to be altered, the bulk of the box stuck out on the offside of the machine where it got in the way of back-set rests and such.

I grilled Bob Burgess at length about this box before I knew what it was because I spotted it in a publicity picture of a 1931 or so KTT — another of those faked, retouched pre-Show photos which lead historians astray — and it turned out Bob had once used the box, strictly a one-off, in one of his own machines. And found the snag. The high kick-start ratio made starting impossible in winter when oil congealed in the camshaft drive. He had to play a blowlamp on the vertical shaft to get a start one day.

Flying start to stardom. Freddie Frith in a club gymkhana on his first cammy Velo

Fitting a heavier mainshaft presented a knotty problem. With Velocette's utilization policy it was unthinkable to make a new clutch or even a new series of final drive sprockets. Eugene overcame it by reducing the shaft to normal size where it ran in the sleeve gear and fitting a split bush so that it could be assembled. A small point but indicative of the constant desire to alter as little as possible at one time.

But the "Eugene" box, although a one-off and soon abandoned, did play an important part in the development of the Velocette gearbox for its selector forks were operated for the first time in Velocette history by a rotary cam plate. The cam plate was large and necessitated an awkward bulge on the front of the already bulky gearbox shell. The operating lever for hand or foot operation was fitted to the cam plate spindle and because it travelled through a wide arc it suffered from lack of travel in the operating linkage at the extremes of movement. When the next gearbox was evolved around 1932, the rotary cam plate was retained and operation improved by linking the operating arm to the cam plate with a striking mechanism, so that a small arc of movement produced a greater rotation of the cam plate, which could then be made smaller. This new gearbox, a team effort in which all the best ideas of previous boxes plus some up-to-date improvements such as the Luce ratchet wheel kick-start mechanism with positive withdrawal of the wheels by means of an external ramp acting upon the folding kick-start crank, was a constant-mesh design (the Willis four-speed was a crash box). At first it was arranged for hand or foot operation, in the latter case a slightly modified Willis drum mechanism being used, but then a very clever redesign combined the positive-stop mechanism with the internal cam plate mechanism. All that was then necessary outside was the box was a simple operating lever; and for many years - until the 50s - it was a light tubular lever.

It pivoted in a bush in a lug on the top of the gearbox end cover and was adjustable both for reach and height. This with changes of internal ratios became the standard box right through the range for some 20 years and was still the basis of the very last gearbox. It is amusing to note that when fitted to racing machines where different pedal arrangements were used the lug was still left on the top of the cover. This was typical of the Willis approach to race tuning. He did not believe in wasting time drilling holes, cutting off lugs and hollowing out bolts to save ounces. He based his work on the simple premise that if a machine was "designed right and put together right it would go right". Adding as a rule: "And if it's going right leave the b..... thing alone."

The cam plate and positive-stop mechanism was, of course mounted at the front of the box, but when the new cradle frame was designed for the Mk. V KTT, the MSS and the Mk. II range of ohc models, with its distinctive vertical seat tube behind the gearbox, the whole mechanism was transferred to the rear of the housing and the pedal linkage reversed. The two-stroke GTP inherited the new four-speed gearbox in slightly lighter form in 1933, which went a long way to making it the most charming two-stroke lightweight of all time.

There may have been a couple of false starts in the development of the four-speed box, but it was third time lucky, for the production four-speed Velocette gearbox is widely admitted to be one of the best designs that has ever been put into a motorcycle for an enthusiast's delight. The vintage years are now so far behind us that it is possible to take an objective view of the influence on the sport and national prestige of an individual manufacturer. I would say that the contribution of Velocette in this respect was considerable and out of all proportion to the size and capital of the firm. We have already seen how a handful of mildly tuned but otherwise standard two-strokes made their mark at home and abroad. From the moment the K was introduced there was a ready-made clubman's machine to help the enthusiast on to the first rung of the competition ladder. A machine which, by its stamina allied to a good performance, enabled the private owner to compete with a good chance of success and few repair bills.

Many riders later to become famous started with a KTT Velocette.

Because of the excellence of this early model Freddie Frith got his name down early for a KTT and entered for the first Manx Grand Prix in 1930 (prior to this the race was called the Amateur TT). He finished third, one of six Velo riders in the first six places, and had the cheek to get up to third place on the fourth lap of the Senior before striking trouble.

That is one story. It is echoed a 100 or more times in the histories of other riders who started a successful racing career on Velocettes. Without fear of contradiction, I will say that no other make of machine in the world put so many riders on the right line for a road-racing career, inspired them to compete, and give them the confidence to continue. Certainly no other machine brought to the Island so many riders from the Commonwealth, riders of strange tongues and colours who had one thing in common. To ride. An irresistible urge to visit the Mecca of speedmen... and a black and gold Velocette. It came about because agents throughout the world fell for the looks and the reputation of the KTT, were infused by a competitive spirit, and ordered one. If they couldn't ride it themselves there was always some local lad they could sponsor. They would jolly well show the rival agents what their bike could do.

The KDT - Kammy Dirt Track

AS FAR as can be ascertained, only one example of the dirt track model KDT survived. It was discovered quite close to the Hall Green works and had been modified for grass-track racing by the substitution of girder forks, ex-Norton, a standard gearbox with a rudimentary foot change and a roadster back wheel. Despite this disguise it was easily recognisable as a DT model by reason of its special frame, quite unlike any other vintage model. It came into the possession of Bert Milnes, a Leicester dealer and vintage enthusiast, to keep company with his vintage cammy trials model and a brace of Mk. 8s. The engine proved to be a normal 348 cc cast-iron K model, not the oversize dope motor. To give his mechanic Rex Caunt a ride in a Vintage Clubs Cadwell meeting, Webb girder forks were fitted and although, with the steep head angle and back set, low-slung engine, it looked a most unlikely road racer, I am assured that in fact it handled quite well. Having spotted a set of speedway Webb plunger forks in the 'graveyard' of another dealer and vintage enthusiast, W.H. Balderstone of Peterborough, I prevailed upon him to part with the forks in a good cause. A photograph shows the machine looking pretty near standard save for the roadster rear wheel.

The general layout will be obvious from the photograph but what you cannot see is the peculiar arrangement of the lower pair of chain, or torque, stays. The nearside stay is appreciably — a good inch — higher than its offside partner. This is to keep the stay out of the way of the chain while securing the maximum triangulation on the offside. I have heard that the first models were built with the contemporary horizontal chain stays but whipped so much that the chain came off. This reborn Velocette, a most interesting sideline of Velocette history, has not yet had a proper try out on a speedway — a lap High Beech does not really count — but as the Vintage Club does occasionally have vintage speedway races by courtesy of friendly, and amused, promoters it may do some day. Vintage Velo enthusiast Jeff Clew has what is believed to be an original big-bore dirt-track engine now in a replica frame.

A very rare photograph of a rare model; the KDT speedway Velocette. Hubert Clayton is the rider

The 'Ugly Duckling' gearbox

WHEN I described the second experimental four-speed gearbox, the one attributed to Eugene Goodman, I did not know that the actual box was in the workshop of a friend at that point in time! Had I known, of course, my task would have been much easier. I had first got on to this link in the chain of development of the late production gearbox through spotting an oddity on a retouched "works" photo of an early 30s KTT. A gearbox which, despite the disguising touches of the artist's brush, did not look quite right. There was, for instance, a semi-circular growth at the rear suggestive of a conventional segment and pinion kick-start mechanism. Not at all in the Velocette tradition of keeping the kick-start mechanism unobtrusively concentric with the layshaft and up to that time avoiding the brutal engagement of ratchets. There was also something odd emerging from the front top of the end cover which defied recognition. It might have been an arm or a pipe or anything. Bob Burgess came to the rescue for although he did not remember this ugly duckling of a gearbox being fitted to a KTT, he remembered the box well for, as previously related, he had actually used it. From his detailed description of it I formed a mental picture and did my best.

The next thing that happened was that Eric Thompson, then Secretary of the Vintage Club, handed me a mysterious cardboard box.

"Have a look inside when you have time, it might interest you", he said. When I did, my eyes nearly popped out.

There, complete save for some odds and ends, was the "ugly duckling". Study of it confirmed the Burgess recollections in entirety, even to the split phosphor-bronze sleeve gear bush which enabled the oversize mainshaft to be assembled in the standard sleeve gear. The accompanying photo shows the sturdy gears and relatively large diameter of the rotary cam plate and its housing, and the considerable overhang on the offside of the machine. No kick-start mechanism remains although the bush for the crank shaft is visible. It may well be that this box, with the kickstart crank removed, was the one used for the mock-up KTT picture. The arm at the top of the outer cover now emerges clearly as a conventional hinged clutch-operating arm, the type of crudity which has excoriated ankles since the beginning of gearboxes with clutches. It operated a push rod which ran through the casing to a conventional Velocette hinged clutch thrust bearing turned through 180 degrees, thus eliminating the internal and therefore mysterious Velocette bell crank mechanism.

It seems the Eric Thompson came by this gearbox with a miscellany of vintage Velo spares from an enthusiast in the North. It had been used but not, apparently, very much.

The Eugene Goodman-designed four-speed gearbox

Velocettes supercharged a KTT and Willis called her 'Whiffling Clara'

They started by hitching a vacuum cleaner to a KTT engine in the test shop

The demand for the KTT and all cammy models which directly followed the TT successes stretched the limited resources of the factory and may well have been responsible for a decline in their racing fortunes. Certainly apart from a few one-off experiments they came up with nothing new in 1930 and 1931, and Willis was the only works rider; being a works employee he did not have to be paid a retainer. Meanwhile, other firms, stung by the indignity of being trounced by a small family firm and with abundant evidence that success in the Island produced sales, redoubled their efforts and dug deep into their coffers to buy brains and riders. First Rudge scored with their grand slam 1-2-3 in the 1930 Junior, next it was Norton who were to dominate.

Reading only the long list of Velocette victories, and seeing perhaps that aerial view of the Hall Green factory, one might imagine that in the vintage years the firm was much bigger than it really was, though I have already said that its contribution to the sport was out of all proportion to its size. In fact, it was still a small family concern having to watch the pennies all the time and ploughing its profits back to buy machinery and sustain its shoe-string (by comparison with others) racing programme. Other firms were going bankrupt on all sides, but Velocettes somehow won through the lean times of depression and unemployment.

One factor which stood them in good stead was their world-wide export market. At times they were exporting 50 per cent of their output. Racing had won this market for them and the KTT model expanded it. But to balance a budget in difficult time, good housewifery is important and that is where Ethel Goodman made a vital contribution. Trained in her father's office in the pre-war days, she took over the Veloce office, did the buying and kept a tight hand on the purse strings between the wars. As a buyer she was as astute as any housewife in a market: shopping around for the keenest price for the hundred and one items that were needed in the production of a machine. She had no technical knowledge, but it was indoctrinate in the firm's policy that materials must be of the best quality; there must be nothing inferior. Apart from that consideration, she was out to get the best buy. At one time she effected a considerable saving by buying steel from Canada. It was up to specification and cost less. Suppliers came to respect her for her keen business sense - for many years she received Christmas boxes from suppliers in the old days. Stock control was doubly important when capital was limited; she introduced one of the first card stock control systems in Birmingham. That was the result of her father's training and his watchful eye on all departments. She remembered well him putting his head round her office door and saying: "You've got 12 carburettors over this month." Twelve carburettors represented perhaps £25, an important sum to Veloce in those days. John Goodman died in 1929. He had seen the word Velocette borne triumphantly around the world, his eldest

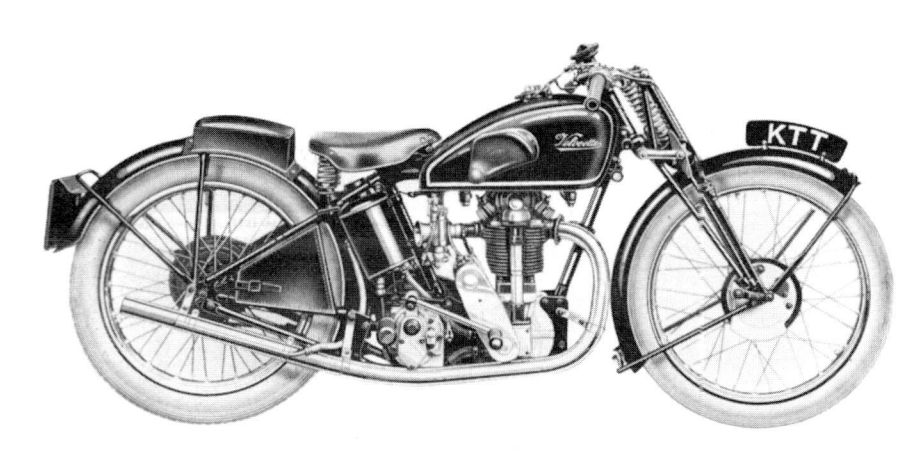

Racer: the 1933/34 Mk. IV KTT

"Whiffling Clara"; is the device under the front of the petrol tank an oil-cooler? It is in the same position as the cooler on (for example) the BSA/Triumph 750 threes of the late sixties?

son carried aloft in triumph as designer of a TT, winner, his younger son and daughter competent in control of their departments of the firm he had founded in a strange land far away from his birthplace.

Ethel kept her hand on the purse strings until after the second war. Pennies were always important. When in the early days of the war the firm was switching to war contract work, but unable to get material, she went right to the top and tackled the Ministry of Supply with the forthrightness she had used with Birmingham merchants. She got what she wanted. More than she wanted because, being used to bargaining, she asked for double the quantity. Percy was with her but not much help.

"He was not much good at that sort of thing." she said. Yet with all the years of day-to-day problems she could still remember when Percy asked for money to pay for his oscilloscope - about £40 it was.

"It seemed a lot of money," she recalled. When at the end of a lifetime of service to Veloce Ltd. she went up to Scotland Yard with her husband, George Denley, who was hoping to sell the LE for police work, and to while away the waiting time watched boats sailing down the Thames, it was, I feel, a symbolic ending to a chapter in this human saga.

"Zoom in to lone figure looking across water — traverse to boats on water — fade out."

Although the Willis experimental long-stroke engine was discarded after the TT (not actually discarded, because it was later fitted with an 80 mm barrel — ex speedway engine, I should think — to make an oversize roadster for a friend of Willis, the first '500' cammy) because it was considered that it did not offer any marked advantage over the normal bore and stroke, it was certainly no slouch. Willis was fourth on the first two laps and third on laps three and four when he went out with unspecified mechanical trouble. But the "foreign invasion" put up a good show for newcomers to the Island on over-the-counter machines. Don Hall, from South Africa, who, of course, did know his way round, having finished sixth in 1929, actually finished fourth behind the all-conquering Rudge trio of Tyrell Smith, Nott and Graham Walker. A.G. Mitchell, who, like his brother H. Mitchell, was having his second shot at the Junior as a clubman enthusiast, was a creditable seventh; his brother retired. J.G. Lind, from South Africa, was 11th (the record book says it was on an AJS, but my information says Velocette), T. Oscarrson of Sweden was 13th and Japan's Kenzo Tada 15th, all these getting replicas. Tada was the happiest man in the Island. Somerville Sikes, an undergraduate enthusiast better known for his abortive attempt to ride a blown Ariel Four in the 1931 Senior, finished 22nd.

Velocette policy remained the same for 1931. While Nortons and Rudges fought an expensive pitched battle for supremacy, using professional riders, they were content to support a number of private owners on slightly improved KTT models.

The improvements were directed at improved reliability

rather than more speed. The thick base "Brooklands" barrel was now standard. The connecting rod, big end and flywheels assembly was strengthened; Velocettes were now making their own big ends. To reduce wear and tear on the cams the base circle was reduced and yet another design of rocker was introduced to obviate the occasional breakages which had always dogged the camshaft engine. The only change in the bicycle concerned the brakes, which now had ribbed, malleable-iron drums.

There had been rumours early on that Velocette would field a 250 — it would have been easy enough to make one out of the experimental long-stroke unit — but the firm issued a denial. (The rumour may have started because they were in fact thinking about making a 250 ohv roadster.) But Willis gave the newsmen and the pundits something to talk about for he turned up with a supercharged model. According to Phil Irving, who, after hitching a ride from Australia with a globe-trotting motorcyclist, got himself a job at Velos, the supercharged model started in the unlikely way that many good and bad ideas start — over pints in the local.

For the first suck-it-and-see experiment Willis hitched up a vacuum cleaner to a KTT engine in the test shop and let it blow into the carburettor. A gallon can was interposed as a balance chamber. This rig seems to have worked beyond expectations. Irving recalls bhp jumped from the normal 24 to an outrageous 32, so the idea was put on a proper footing. A small rotating vane blower (Foxwell was the make, though I have never heard of it before or since in connection with any other supercharging experiment) was mounted in front of the engine where dynamos were usually fitted and was chain driven from the engine shaft. The chain was enclosed in a case which looks very much like the dynamo belt cover pressing for the GTP model. The blower blew air into a pressure chamber mounted where the oil tank usually was. It discharged into the mouth of the carburettor. Now this system of blowing air into the mouth of the carburettor instead of compressing petrol vapour from the carburettor is the hard way to supercharge. Admittedly Mercedes always used this layout, but most people adopted the other way because if you blow into the carburettor you have to pressure balance the float chamber and the petrol tank or you will have fuel spurting out all over the place.

Willis-ese: A valve was a nail, a piston a cork, the test shop a din house...

HAROLD WILLIS has made so many appearances on and off stage in this Saga I have tried to dramatize that many will feel that they have built up a mind's eye picture of Willis the man. If so they may be wide of the mark. He will always be remembered most for his Willisisms which, handed down through the years, became an essential part of motorcycling phraseology. Some like "double knocker" for twin overhead camshafts have become so hallowed by common usage that quotation marks are no longer obligatory. Because of this streak of whimsy which demanded that nothing was ever referred to by its correct title — that insisted that a valve was a nail, a piston a cork, a horse a hay motor, the test shop a din house — it is too easy to form a picture of Willis as being a light hearted extrovert... the Court Jester of Hall Green. This is

Harold Willis refuels at the end of the fifth lap, 1928 Junior TT

Willis went about it the hard way because he did not like the idea of having a balance chamber of about one gallon capacity full of an explosive mixture. Certainly not when the bang, if any, would occur between the rider's legs. And as he was going to ride the thing, who can blame him. I have never heard of a balance chamber blowing up, but there is, of course, always a first time. I would have thought the idea of a pressurized fuel tank was equally uninviting, but Willis seems to have regarded this as a lesser evil, and a complete machine was built for the Island bristling with balance pipes and taps which had to be opened to release the built-up pressure when the machine was stopped.

It was this pressure build-up which caused Willis to think up the immortal name, "Whiffling Clara". A pressure chamber was considered essential, of course, to store up the gas from the constant-delivery blower in the interval when the inlet valve was shut.

This time the blown Velo was regarded as a viable Junior runner. The unblown TT mounts had two innovations. Hairpin valve springs were used for the first time — under licence from Sunbeams who seem to have the patent angle tied up — and the new constant-mesh cam plate-operated four-speed gearbox was fitted. Positive-stop operation was still provided by the Willis drum unit, fitted externally and modified to cover an extra gear position. With the hairpin valve springs went a redesigned head with one of the new baby 14mm plugs. To get the plug nearer the centre of the combustion chamber the cam box was modified. The KTT plunger cam box scavenge pump had always got in the way of the plug so it was now replaced by a very slim gear pump on the end of the cam box.

Cast-iron was still used for the head, though other makers were beginning to use bronze. The bicycle parts were as before except that the tank mounting was lowered. Incidentally, Clara, who had a great thirst, had a pistol grip tank probably inspired by the TT New Imperials.

Something of a pre-race mystery was caused by the entry by Velocettes of a mystery rider, A.N. Other, reported to be a famous TT rider. It did not need much deduction to arrive at Alec Bennett as the man most likely to come out of retirement for the sake of his old love, and so it was. There were the usual Velocette enthusiasts in the lists. Alec wide of the mark. Humorous he certainly was but in an unsmiling way: his humour so dry as to be astringent. Rarely did he let his feelings show. Some of this may have resulted from an early naval training at Osborne which pitchforked him into the first World War as a midshipman to be torpedoed and rescued from drowning. His love for motorcycles compelled him to spurn the security of his father's business or the fruits of his general engineering training, but at Hall Green with the Goodmans he found happiness.

As a rider he was tough and reliable if not possessed of the elan of his best friend, Wal Handley. His TT career spanned 1924 to 1932 with two Junior seconds on Velocettes, a fifth and a ninth on Montgomerys, and a 12th on a Triumph.

Only once, as far as I can ascertain, did he fall off seriously in the Island. Although a man of fixed ideas and sceptical of innovations until they had been tried and proven (remember the little matter of the modified tappet adjuster which, rightly as it turned out, he would not adopt until it had been proved), he enjoyed experiment, as witness his work on the early spring frame, the long stroke, and the supercharged Clara... and, of course, his positive-stop footchange. Dogmatic sometimes, when others tried to argue with him, and always resistant to change for changes sake, he was a perfect sheet anchor to restrain the enthusiasm of other members of the Hall Green crew. Only in his spare time did he unbend enough for others to get some insight into the real Willis behind the enigmatic mask. He would climb into his very second-hand aeroplane, "Clattering Kate", and fly to the Welsh coast where he had a cottage and a home-made speedboat. The locals thought the world of this engineering wizard who came out of the sky to be one with them. "He's the only man we know who can understand a three-speed gear", one local once told George Denley. It would be a push-bike hub gear, of course, and the highest tribute they could pay - he is buried there.

"A man's man" is how Mrs. Denley (Ethel Goodman) summed up Harold Willis, who remained a bachelor whose interests unexpectedly included music and ballet. He died tragically from meningitis following a minor sinus operation, just as Stanley Woods was about to win the Junior for Velocette and was trying out the supercharged twin, the Roarer in Willisese, which was the last engine he was to run in his din house.

In a way I am glad he did not live to see British racing supremacy eclipsed. If he had lived, the iconoclast in him would have been sorely tried to find the right words.

Velocette 1926-28-29 TT Winners

"1066 and all that" is how Harold Willis referred to this tank transfer. Veloce never won a TT while their machines carried the transfer. The Vintage MCC stocks these and ones suitable for almost every British bike made

Mitchell, Wilf Harding, Brewster from Australia, Fondu from Belgium, Meageen, Jimmy Lind (down to ride Clara in the Senior as well), and new boy L.G. Archer, who had been assisting Willis in trying out the new models at Brooklands. Jack Williams was down with Bennett and Willis as works entries. There were even more Velos on the starting grid for Francesco Franconi and Renier opted for them when the French Jonghis on which they had entered were scratched after a fire destroyed the factory.

With the usual Rudge-Norton battle raging from the start Velocette were barely in the picture on the first lap, with Meageen 10th and Franconi 12th. As the race wore on reliability paid as more and more Velocettes came on the leader board, Meageen getting up to seventh before dropping out on the third lap. Willis went out early after stopping to fiddle with Clara on the Mountain, but a rising star was in the ascendancy. Young Les Archer was coming up fast. On the third lap he was eighth ahead of Mitchell, who was 10th. Pit stops reversed these positions, but by lap five Archer was seventh, the leading Velocette, just ahead of the wily Alec Bennett, who had suddenly come from nowhere. At the finish Archer was sixth behind Graham Walker's Rudge, and next came Bennett, level with Les Davenport. Jack Williams was 10th, gaining the last of the replicas, and there was a fair sprinkling of Velos among the remaining finishers. Lind, it transpired, had gone out with a broken frame and this added to the conviction of several riders that the old frame, still basically the original 1926 design, was getting past it. The hairpin engines were turning over faster, too, and there were complaints about vibration. Willis attributed his retirement to a "camshaft weevil" which, being interpreted, meant a broken rocker.

Clara really showed her paces in the Senior for Lind was 10th on the second lap and 11th on the third when he was eliminated by the silliest thing, the loss of the jet cap of his carburettor, but Franconi from Switzerland kept going to finish 11th, the only 350 finisher, and was awarded the Visitor's Cup.

The lesson of the TT were soon translated into production modifications of the KTT: the Veloce policy of improving the breed by racing. A tubular sub-frame running from the bottom of the crankcase to the rear fork and braced to the gearbox lug by plates which provided a convenient fixing point for footrest and pedals reinforced the frame. A larger, lower tank was fitted and the new type four-speed box now fitted as standard added up to the Mk. IV KTT, one of the most handsome of the line. Early models had cast-iron heads like the TT models, but a bronze head was soon put into production and this permitted high compression ratios. Rapid wear of the valve seats — there were no valve inserts — was reduced by allowing the valve to rotate.

This new KTT model, the Mk. IV, was the first to bear an official Mk. number. I say official for the previous 1931 onwards model with thick-base barrel, small diameter cam, improved big end and cast brake drums had been known in the works as the Mk. I to distinguish it from earlier models. Unofficially because George Denley as Sales Manager did not believe in mark designations. A KTT was a KTT and that was that, was his ruling, which was not popular with the spares and service department. As this difference of opinion has become something of a minor Velocette legend I asked him about it. He was perfectly frank.

"My job was sales and I had to think of the dealers. When you bring out a new Mark number it makes the previous machines obsolete and difficult to sell. If I had been on the service side it would have been a different matter. I should then have been in favour of Mark numbers."

At this stage I feel it worthwhile to go into the question: What was Clara like to ride? The man who knew better than anyone, having been the last one to use it, was very much alive at the time I was researching and writing this section. Les Archer, of Archers of Aldershot, L.J., to use initials for identification, earned the nickname "The Aldershot Flyer" the hard way in the 30s, first at Brooklands and then in the Island and the major Continental capers. He started his Brooklands career on New Imperials, but, like so many other racing men, he did not really shine until he got astride a Velocette. After that "his life changed" as they say in novels and the names of Archer and Velocette became almost synonymous. He told me something of his "love affair" with Clara. From the start, and always after, he seems to have cherished her with the tenderness one would show to your best friend's girl for Clara was Harold Willis's own special pet and Les felt greatly honoured when Willis put her in his care.

Certainly Willis had allowed Lind to take her out in the Senior, but this was not quite the same as letting her out of his sight. This seems to have been the underlying reason for his celebrated retirement in the Ulster, one of those Ulsters when the heavens opened, everyone was slowed to a crawl, and the canny ones pulled in and complained of water in the magneto or in the carburettor. Archer brought Clara in and scorning the mild deceits, spoke the words which became immortal in Ulster legend: "I am not a fish". Clara, to her credit, was not put out by the downpour and the thought of Willis's dearly beloved being cast away in the appalling conditions was too much for Les.

When you know something of the problems involved in riding the blown job on a road circuit you will understand his reluctance to risk it.

The cockpit drill was very complicated. All supercharged machines of the period suffered to varying degrees from "indigestion". I know the symptoms because I own one. They have the tendency to keep going when you shut off and to refuse to open up afterwards... at least when you want them to. To try and overcome this trait Willis had, in the Mk. II version, done quite a lot of original work. There were two throttle controls linked together, one on the carburettor which cut off the supply to the blower, and the other (merely a carburettor less float chamber and jet assembly) between the pressure chamber and the engine. The reasoning was logical, although I do not remember anyone else trying it. There would be a lot of gas in the blower and the pressure chamber and piping, but the second throttle would stop it getting into the engine when the throttles were shut. I gather that this overcame the usual trouble of motoring on when the throttle was shut, but it did not completely cure the indigestion... the accumulation of over-rich and oil-laden (blowers had to be fed with oil) mixture in the induction system which prevented instant response when the taps were opened. Archer assures me that Clara was quite responsive on easy circuits where it was only necessary to ease off and wind on, but whenever it was a case of shutting right down for a slow one there was this "indigestion".

The Willis answer to this problem was both delightful in its simplicity and unique, to the best of my knowledge. He incorporated a drain valve which could be operated by a trigger lever on the handlebar. If it was a corner approached with a closed throttle, and the engine refused to open up, you pressed the trigger and the excess of gas, oil and what have you (Willis had a couple of words for the stuff that came out, but they are not polite enough for these pages) was discharged on to the ground. Well, not always, for according to the angle of lean and the prevailing wind there was the possibility of it blowing on to the back tyre, which was not helpful. Rather a case of "Do not pull the chain when the train is in the station".

I don't think for a moment that Clara was a serious attempt at a pre-war world-beater. It was a typical Willis exercise, chiefly for his amusement, but there was a certain purpose beyond this, as Archer pointed out. With the extra urge that the blower produced it was a very effective way of testing the KTT components. Submitting them to stresses beyond the realms of atmospheric induction. The remarkable thing is that the engine, a perfectly normal iron KTT, stood up perfectly. There was no trouble with melted pistons, the usual aftermath of forced induction, but Clara was a rare one for melting plugs.

Would even melt mica insulation, said Archer. The effect of supercharging was, he said, to increase the power, not the revs. Made it go like a five-hundred. One would have expected it to perform best on the flat-out outer circuit at Brooklands. Apparently not. It was rather heavy (he thought it must have weighed the best part of 500 lb) and because of its speed — it was reported to have been clocked at 112 on the straight — would have to take a line high enough up on the banking with the big stuff and the big bumps. Apparently when it did land after taking off over the famous bump of the River Wey bridge the forks would whip back until the front wheel hit the blower casing. Not funny. You may wonder how it came about that Archer, this slip of a lad — he was then because regular work-outs on the Brooklands outer circuit on the New Imperial machines his father prepared so skilfully kept his weight steady at nine stone six pounds, the battering of the concrete surface crazed by age being as effective as a skilled masseur — came to be entrusted with such a rare piece of machinery. The fact was that since switching his main effort to Velocettes (he still rode New Imperials in other classes), he had taken over from the Brooklands Velo star spot vacated by Freddie Hicks, who had transferred his allegiance to AJS, then making a desperate last bid (the A.J. Stevens firm, that is) to win back the supreme position in racing they once held. A tragic bid, for after a run of bad luck in the Island Hicks lost his life in the 1931 Senior. Like the majority of Brooklands regulars, the Archers had a workshop at the track and this became the official Velocette "outpost".

Archer remembered vividly his first track Velo. "We went up to Hall Green and there was this beautiful bike all fitted out with track footrests and silencer and loads of Float On Air padding for lying down to it." Front tyres 21 by 2.75, back tyre 21 by 3.00, and a pair of spare wheels with 2.75 tyres for short-distance races. 'Don't mess about with it,' said Willis, 'just learn to ride it properly'.

"We took it down to the track and my father told me to do a few warming-up laps and then try it down the (measured) half-mile. After my New Imperial this Velo was simply marvellous. After a few laps to get used to it I couldn't resist flinging it into the Byfleet banking — the result was a lap at 100.85 mph. My father was furious."

He would be, too, for the element of surprise had escaped by this moment of exuberance and the speed would not be lost on the handicapper. However, at their first outing in a race the Archer Velocette duo got one of the coveted Gold Stars for lapping at over 100 mph in a race. Only two other 350 Gold Stars had then been awarded. His interest

The "dog Kennel" engines

THE 1932 results, though demonstrating Velocette reliability, which almost went without saying, clearly indicated that the original design, improved only in detail since 1926, lacked speed to match the new designs of other manufacturers. One innovation demonstrated by Nortons was the down-draught angled inlet port which clearly offered breathing advantages. The original head design did not lend itself to this modification. Another weakness was the difficulty nigh unto impossibility of keeping oil in the cam box when the rockers had to protrude from slots. With increased engine speeds, hotter cams and stronger valve springs, large quantities of oil were necessary aloft. These requirements resulted in a batch of works racers known as the "dog kennel" engines because of the resemblance of the cam box to the traditional gable end with overhanging eaves of dog kennels. The new look in cam boxes was occasioned by the complete enclosure of the rockers in the overhanging eaves, the valves with exposed hairpin springs being operated via short tappets, as later became general practice with dohc designs like Norton. The bottom end of the motor was reminiscent of normal KTT practice, but a new barrel with through bolts and a new head having more likeness to the M series of push-rod engines gave the engines an unfamiliar appearance. Only 350 editions appeared in 1933 but the following year a 500 version was introduced.

The works 1933 'dog kennel' TT machine

in the Isle of Man started soon afterwards, the direct result of a trip to see the races with local trials and scrambles star Len Heath.

"You know, dad, we shall have to have a go at this road-racing game," he said when he got back, so a KTT was ordered for the 1932 Junior.

"When we got over to the Island we went to the Athol Garage and there were all the models lined up — there must have been 20 or 30 of them — just like in a showroom. Each one had a label on it with the rider's name."

This was the Willis way of ensuring as far as possible that the KTT models built with loving care under his personal supervision, the engines bench tested and tuned, would not be "mucked about" or worn out before the race. If the machines had been delivered to owners beforehand, few would have resisted the temptation to run them in minor races or pull them down to see what was inside. This policy may have had much to do with the remarkable reliability of ordinary KTT models in private hands in the Island.

Archer certainly proved the point that the models were turned out ready to go for he finished sixth ahead of Alec Bennett and was first Velocette home. Archer was to prove a valuable addition to the Velocette ranks for many years to come. Only once did he blot his copy book. It was the next year, when he came off at Ramsey on the first lap of the Junior and put himself out of the Senior as well. He made up for it in 1934 when he took over the "official" Senior mount which was to have been ridden by Gilbert Emery who hurt himself in practice. This was one of the rare 500 cc "dog kennel" rocker-box models on which Walter Rusk was third and Archer fourth. But although Archer was usually around the place in the Island and in a number of Continental races it was at Brooklands where he really made history. And the most historic achievement was winning the Hutchinson 100 in 1933 on a 350 Velocette at over 100 mph — 100.61, to be precise, the first time a 350 had packed one hundred miles into the hour at Brooklands where silencers had to be worn. The Hutchinson Trophy was incidental and he did not expect the handicapper would give him a chance. The main problem was fuel consumption.

To hit the target it was necessary to run through non-stop. The engine, the one on which he had finished sixth in the 1932 Junior, held the lap record at 104, but that was on "blow the consumption" alcohol fuel. The delicate problem was to equate low fuel consumption — his tank held 4¾ gallons — with an over-100 mph lap speed. It was done by raising the compression ratio to 11.2 to 1, yet running on RD 1, which was an alcohol blend intended for much lower ratios. It is one of the finest tributes I know to the stamina of that wonderful Percy Goodman design. Asked later for the secret of this 100 per reliability, Archer's considered opinion was that it was generous oil circulation which cooled Veloce internally. We know now that oil cooling of the internals of a racing engine is an essential requirement. Not every designer appreciated this fully in the early 30s. When Percy Goodman redesigned the model K to give it full pressure dry-sump lubrication in 1924 he was a visionary.

I cannot leave the subject of this epic high-speed enduro without reference to the physical endurance of the then sylph-like Archer. For an agonizing hour he had to lay absolutely flat, chin pressed on the tank top, elbows and knees tucked in, holding the Velo on the tightest line (when not weaving in and out of the "traffic"), subjected to the hammering of tiny tyres through the rigid frame. With but a mile or so per hour to play with, the slightest relaxation to ease cramped muscles would instantly drop the lap speed below the target.

"Once I must have moved my head a little because I was getting cramp... it dropped my lap speed by a couple of miles an hour and my father was waving a signal board at me," he told me. "They had to undo me from the bike at the end," he added.

For good measure that year, Archer won the Brooklands Junior and Senior Grand Prix events — the first time a 350 had won the Senior event — a 50 mile race, the Wakefield Cup for the second year running and, not surprisingly, the Bemsee 350 Aggregate Cup.

No matching numbers!

POVERTY decreed that both the Mk. I cammy models I have owned... and loved... were non-standard assemblies based on cycle parts from the despised KTP model. The first one started my love affair with Velocettes in the early thirties. I can truthfully say it was the first really good motorcycle I owned and that was after owning a machine which collectors drool over, a flat-tank Model 80 Sunbeam, the little brother of the illustrious Model 90. Its numbers matched but despite its pedigree and hairpin valve springs it was a poor performer with a hard ride which made rough roads purgatory.

By comparison the Velo with its Webb forks, comfortable riding position and smooth engine was a dream bike. It came cheap because it was a bitza, a KTP, the dreaded two port coil-ignition model which had been converted to a KSS look-alike with a single port head and a magneto. I knew the previous engineer owner had put it together properly because he was the only man in my part of the world who knew how to adjust a Velo clutch and was not afraid of hunting-tooth bevels. I was mortally afraid after hearing tales of folk who had light-heartedly disturbed them, never to get the timing right again. So afraid that when I did eventually risk taking the head off for a decoke (and to look inside), I jammed the engine with a wooden wedge behind the sprocket so it could not turn, and likewise jammed the top bevel. Careful not to disturb anything while I attended to the valves I was still most relieved when it went back together and with the wedges removed actually ran. You may think this is altogether childish but you must realise that in those days there were no workshop manuals, no DIY step-by-step periodicals, not even a maker's hand book if you bought second hand. In fact I never saw a Velo handbook until I had my second KTP bitza in the 50s.

In nearly two years of ownership the Velo took me to work and to parts of the country to which I had never dared venture before, never let me down and gave me a new appreciation of what a motorcycle could be. Bear with me but I can still remember as yesterday how, with a friend on the pillion and a tent and survival gear hung all round we set off for the unknown lands beyond the Midland counties and encountered people who to us spoke with strange accents. We proved to our satisfaction that if you went beyond Lands End you would fall in the sea before making our way along the South Coast as far as Portsmouth whence we headed for London.

Neither of us had been to London and we chickened out when we reached the seriously built-up part. Besides, the money was running out. The clutch began to drag and fiddling with the cable adjuster didn't help much but eventually resulted in a mysterious screech from the gearbox region. Knowing no better and nothing at all about bell cranks and "gates", I poured more oil into the gearbox and before long all was quiet again though the clutch still dragged. I suppose overfilling the box resulted in some oil getting to the tortured thrust race. Starting was no problem. There was no kick-start because of the KTT type back-set footrests; anyway it was much more manly to run and bump start.

Why did I part with this paragon among bikes? Well, when the second winter came round I found it almost impossible to start on a cold morning. The 50 grade straight oil (multi-grades had not been invented) locked a cammy Velo virtually solid on a winter's morn and the only thing to do was to play a blow lamp on the camshaft side till it freed. I hadn't a blow lamp so a kettle of boiling water had to do. And the gas lights (no dynamo) left much to be desired (I used a whole box of matches between Peterborough and Leicester one blustery night). And there was this girl whose parents would not allow her to ride pillion.

The second KTP bitza came about because my son, having cut his teeth sprinting, wanted to go Vintage racing and in the sixties the Vintage Club was short of racers. The bulk of it came from my good friend Ivan Rhodes in a part swap deal with a vintage AJS, the KTP element of frame, forks and tank was largely due to such parts still having little market value in those days. Son Roger, apprenticed to the motor trade, probably learned more from building up the assorted parts and drilling lightening holes than on day release at the 'tech'. The available wheels were 21in front and rear and I produced a brand new pair of 3.00in Dunlop Universals (war grade) from my stock of war surplus tyres. Believe it or not, those same tyres are still in use as I write in 1994 and no one who has ridden the machine has complained about the lack of grip. Perhaps they have been too polite but I think the large diameter has a lot to do with it. After a season or two the KTP-KSS became my road racer and sprinter and still is one of my most cherished bikes. We've had some exciting times together, races at Oulton Park, Cadwell, Mallory many times, sprints all over the place and the odd hill-climb (won the Vintage

Titch Allen on the bike that looks like a KTT, but isn't, at Oulton Park. He says it's the only known shot of him really trying! The other chap, on a similar Velo, is trying too

class at Hoghton Towers once), and most of the work done on it has been by way of updating to make it more like the vintage enthusiast's dream, a Mk. I KTT, and that has been due to good fortune and not a little help from my friends. The first bit of good fortune was many years ago when the Geeson brothers, they of the delightful do-it-themselves museum in Lincolnshire, owned and loaned for racing a 1932 or thereabouts KSS which by some mischance had a pair of strutted Webb forks.

As they were concerned about originality I felt it was my duty to offer to exchange my plain Webbs for their strutted ones! Nowhere could I find one of the three-pronged click-action filler cap for my KTP tank so I made a tower to take a winged alloy cap I came by and it looks like a KTT job.

Fortunately the front brake plate was an alloy one from a KSS or KTS, not the KTP pressed steel one. It came cheap because someone had ruined its originality by drilling holes in the water drain flange. Wear of the cams and skids is a problem with these engines if revved a lot or run with strong springs or cheap oil so the acquisition of a cam box scavenge pump for a KTT enabled me to take a feed from the bevel housing inspection plug over to a jet squirting down on the cams and followers, the extra pump taking away the extra oil. The type 6 Amal I used at first was replaced with a TT AMAC instrument which has no throttle needle and was used on most racing machines up to 1928, a very easy carburettor to tune having an adjustable fuel control for the pilot mixture like later TT carbs and an external air slide. The best bit of good fortune came when I bought a job lot of early cammy Velo parts and what I took to be at first glance into the tea chest as a normal three-speed gearbox turned out to be a four-speeder. I don't think it's "ex-works" but more likely to be the work of an engineer Velo enthusiast as it is basically a 1933 type box with the end cover like the three-speed box. There is no provision for a kick start and my guess is it was built up to keep it slim enough to allow the normal KTT rear-set fixing arrangement. The positive-stop is the Willis drum type and it has transformed my mongrel as, let's face it, the old three-speed box is not the vintage Velo's best feature, being directly descended from an early two-stroke design. So that's why my favourite Velo looks a bit like a KTT and not being in the money market I don't think I would swap it for a genuine one. Even if the numbers matched!

Far from being ashamed to be using a KTP frame I have always found it had certain advantages (apart from price) because it is much lighter and the tank fitting, by means of ears front and rear to bosses on the frame, makes tank fitting easier and allows one to fit a smaller tank for speed work The discovery in the post-war years that it was a stronger frame because it had a steel gearbox lug instead of the fragile malleable iron one on other models pleases me immensely. Velocettes must have thought so because they adopted as standard in 1932.

While it is confession time I had better own up that I used yet another KTP frame and tank on my Mk. IV type KTT which I used more as a road bike than a racer.

It began as a genuine KTT bought and raced by one Paddy Cash who was a grass track champion in the South, but after his death it was used as a road bike by his tuner, Mike Erskine.

Wanting something more civilised with lights and kick-start, he eventually fitted the engine into a rigid MSS, leaving the cycle parts to weather outside. A friend rescued them but 30 years of exposure had ruined the frame and most else apart from the subframe which had always been well anointed with oil. The Mk. IV engine was lying around too, having eventually been cut out of the MSS frame with a 'torch'. All this, I assure you, is typical of the unconcern for old bikes before the Vintage Club changed the scene. The remains were given to me for a small consideration (not money) and work began. Ivan Rhodes produced yet another KTP frame and tank (he only builds originals) and he knew a man who had a Mk. IV back wheel but wanted one for a Scott (easy), and the late John Griffith gave me a KTT front hub (it turned out to be Mk. V but no matter), and the same friend who had started this saga stumbled on a pair of braced Webbs in a scrap yard (you wouldn't believe it, would you) and gave me them for another consideration.

Owing to the old Velo custom of oil draining into the engine when left for a period, the engine was full of oil and perfectly preserved; not so the oil tank which had rusted through, but thanks to the modern wonder of glass fibre and resin it will never rust again. Because the engine had been grass-tracked it was not altogether surprising that it had been bored to 80 mm to take a speedway JAP piston although the head was a standard size iron KSS head, which at least provided a degree of squish. An old grass track custom, of course, though I'm not suggesting that anyone had been cheating by entering it as a 350 instead of a 407.

It so happened that Ivan the Generous happened to have a bronze Mk. IV head which had been opened out to 80 mm which was no use to him, so my Mk. IV became a 407, a nice low compression one ideal for road work, and not serious competition. Quite the nicest cammy Velo I have ever had, it turned out to be, with exceptional steering and road holding. The actual build was done for me by Velo enthusiast Mike Vangucci as I was otherwise engaged at the time, and recently, when I realised I had need for only one vintage cammy, I gave him first refusal at a special price and am glad he has been able to give it a good home. Morally, it was part his anyway, I thought. He's not bothered about originality either, if the end result is good to ride. When our vintage bikes were new we could not wait to alter them to suit ourselves and took pride in modifications. Nowadays the alterations from standard which are part of the history of a machine have to be stripped off in the pursuit of the current craze for originality and, the end product, in my opinion, is bland and anonymous.

The restorers will have a difficult job if they ever try to restore any of my bikes!

The four-speed — three-speed gearbox

The revered GTP. And the KTP
So many rude things have been said about the cammy...

Just as the overhead-camshaft models stole the limelight in 1926 and made such demands on factory resources that two-stroke production was halted for a year or more, so have they occupied the recent pages of this story. I have dealt with the "in-between" models, the U and USS, on reflection perhaps a little unkindly, but in this respect I was drawing on my own recollections of the motorcycle scene in the 1927-29 period when the two-stroke was demoted to a Cinderella role. A utility machine for the impecunious of whom there were a great many, myself included — a beginner's mount for teenagers lucky enough to have well-to-do parents and a rode-to-work machine for the more successful artisans. Not to be regarded as a sports machine, although many firms struggling to keep their two-stroke range alive still supported trials. Even the Villiers range, which had gained a sporting reputation with the 172 cc TT Super Sports engine (a title based on not very outstanding performances in the Ultra-Lightweight class in the Isle of Man in the early 20s and a number of class successes in obscure overseas events), had become distinctly utilitarian and the hot-stuff dope Brooklands engine finally discontinued.

The model U fitted into this background more than adequately; only the USS seemed to be a misnomer. However, Velocettes, to their credit, did not let the two-stroke go downhill for want of development and take the easy path of price cutting which usually led to extinction. Instead they did a thorough redesign, setting out to produce a better job than had gone before at a keen price which was obtained by design rather than cheese-paring. The frame was an example of this. By cutting out the loop under the engine they reduced the frame to the practical minimum, the rear engine plates were pierced to house a dynamo, first of the traditional Velocette belt drives, and coil ignition permitted a further saving.

Electric lighting equipment had become a necessary

Give a Manxman a motorcycle and he will use it in competition... Dennis Corkill gives his GTP, standard except for lack of silencers, the flat-out treatment at the Ballamooar quarter-mile sprint near Jurby. Dennis, Chairman of the Manx Motor Cycle Club, finished third in this handicap event, behind an MSS and a KTT

sales point by this time and coil ignition was a corollary on a low-budget machine. Coil ignition soon became a dirty word in motorcycle circles because of badly designed cut-price lash-ups, but Velocettes made a good job of it, housing the contact-breaker on the crankcase and the coil upside down in a cavity in the petrol tank where it was safe and dry, and I do not recall any criticism of the system as used on the GTP. The GTP engine was conventional after the unconventional offset engines before it. I have already suggested that the poppet-valve racing engine was its true progenitor, and I see no reason to change my view. The poppet engine was the only Velocette experimental engine to have a conventional two-bearing crankshaft, was the first to have a roller big-end, and there is a discernible family resemblance between it and the GTP. (There are various theories about the origin of the title GTP, none of which seem to hold up, and I leave it at its literal interpretation G Two Port.)

Quite the nicest feature of the new engine was the oil pump. As always, Velocette made lubrication their first thought, not an afterthought. And, as usual, they used a reciprocating valveless pump not unlike earlier designs but this time housed on the offside crankcase in a housing shared with the contact breaker. Finger adjustment was by an accessible knurled knob at the rear of the housing.

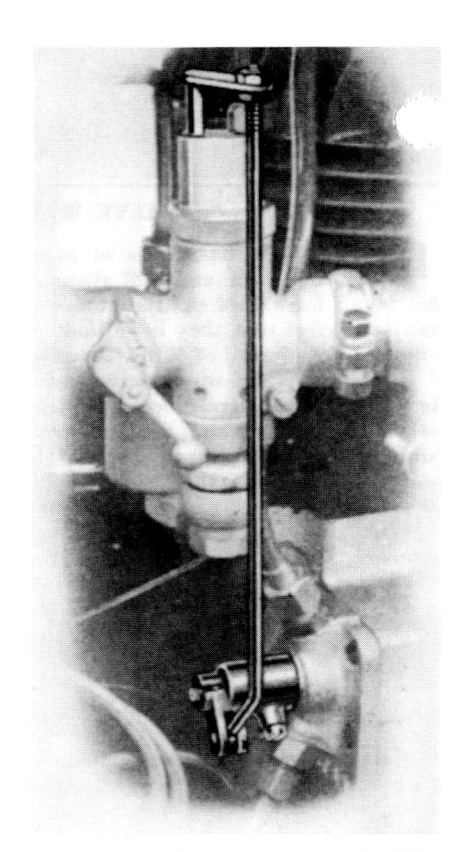

How the first "Autolube" was arranged on the GTP model

Whether this was arranged thus on purpose so that it would be underneath the carburettor and later permit easy linkage to the throttle I do not know... I doubt it, actually, but it was neat and out of the way of road filth.

Announced at the same time as the neat little GTP was another two-port coil ignition model, the KTP. So many rude things have been said about this unfortunate model in the past (the very mention of KTP makes the dedicated vintage Velocette blanch) that there is no need for me to add to them. Instead I will temper criticism with explanation if not justification. True, it had the worst performance of any of its stablemates, but that is an unfair comparison because it was not intended as a sports model. For the reasons for its ill-conceived birth it is necessary to consider the condition of the market in 1928-29. A price war had already eliminated a lot of the smaller manufacturers who, producing in penny numbers, could not get down to the prices set by the big producers like Ariel, B.S.A., Royal Enfield and Triumph. Electric lighting included as standard equipment was becoming a big sales point and the cheapest way to provide

Jeff Clew, doyen of motorcycle historians, with wife Audrey in 1952 at a Sunbeam MCC rally. He had already become a Velocette enthusiast, and so had Audrey, who shared his KTP at early Vintage Club Royston Hill Climbs by swapping number plates. The bike is not a USS, as a Velo fan might think, but a model 32 (£32,) a bargain buy promoted to use up stocks of U parts before the launch of the GTP. Jeff has one of the few USS models in existence. He rode a Mk. II KSS spring frame hybrid when serving as a travelling marshal at Silverstone and Snetterton

First publicity picture of the 1930 KTP two-port model. The artist slipped up a little over "VELOCETTE MADE IN ENG" on the dynamo chaincase

it was by using coil ignition, saving the cost of a magneto. Fashion was playing an increasing part in sales and the arbiters of fashion, notably B.S.A. and Ariel, had plumped for the twin-port exhaust system. This offered a nice symmetrical appearance and the promise of a quieter exhaust — usually quickly overcome by a proud owner degutting the silencers and fitting big copper tail pipes in lieu of fishtails. Except at Brooklands, where as you had to have an expansion chamber and fishtails related to the engine capacity and it was more convenient to have two small ones instead of one whopper, there was no power advantage in twin ports; usually the reverse, although the twin-port JAP engine developed at Brooklands had upheld the fashion in the Island. So ask yourself what you would have done if the sales department and the dealers said they were losing out because you had not a low-price twin-port electrically equipped job in the range. You would probably have done what Velocette did.

Modify the head casting to provide an extra outlet, make a lefthand exhaust pipe and silencer... you already had the offside system. Make a new magneto chaincase — sorry, dynamo chaincase — bigger at the bottom so that the dynamo could run at higher speed, the two-to-one reduction for the magneto was an integral feature of the design. Obvious place for the coil ignition contact breaker was one the end of the camshaft close to the coil which you put in a hole in the tank as on the GTP. You even saved on the wiring. Save cost everywhere else: a steel case for the primary chain instead of an oil bath, and a new frame which had eliminated some of the more expensive lugs. The tank was provided with tabs which bolted direct to the frame, for instance. No matter that the twin-port exhaust took the fine edge off the motor — it wasn't a sports job, anyway. Just the job for the promenade boys of the day.

Now it is one of the facts of life that when you establish your business on the fine principles of honesty and integrity

The model U (left) and its 'Super Sport development, the USS

The 350 cc side-valve Velocette built experimentally in 1931-32 and discarded after tests which showed a lack of performance. In this picture the engine is housed in a GTP two-stroke frame; it was later fitted in the frame used as the basis of the MOV 250 cc ohv model

the slightest deviation from these standards sticks out like a sore thumb and is doomed to failure. For the first time Velocettes had swayed from their principles of producing a functional machine with no frills or furbelows, and the market sensed it. The enthusiasts, of course, spotted the deviation right away and the floating market of fashion-mongers was not greatly tempted either. If another firm with less tradition had made something similar it might have sold, but I suppose they would have tricked it out with chrome and colour anyway.

In fact the KTP was not a bad bike at all. I remember the first and only one that came into my home town. It was bought by an overseas student, which may or may not prove something.

It sounded rather flat, which he soon cured by removing all the baffles and opening the fishtails to a ½in slot. It then made a dull booming sound. I don't remember him having any mechanical trouble, but from time to time oil leaked from the cam box into the contact-breaker and made starting difficult. We turned up our noses at it, but it was a far better machine than any of us aspired to.

I asked Charles Udall how it came about that Velocettes produced the KTP.

"It wasn't designed, it just happened. It was done at the insistence of the sales department. The performance was ruined because the 1¾in pipe on the other side spoiled the extractor action of the exhaust (the cammy Velo always had a very nice trumpet-shape exhaust port in the head which, with the correct pipe and silencer, gave a tuned system). Smaller pipes would have been better, but they already had the offside pipe. It was not a very good machine."

But in one particular the KTP was better than the contemporary KSS models, as vintage enthusiasts are now learning. The frame, though designed to cut cost, was stronger because it had a new steel gearbox lug as used on the KTT. The old malleable lugs were prone to breakage after long and hectic use and eventually the new frame was standardized for the roadster K range. So the best vintage frame you can get happens to be from a KTP. The majority of the KTP models sold were eventually converted into magneto K models by enthusiasts, an original KTP must now be one of the most difficult vintage Velos to find.

Looking back on the early 30s it does appear that Velocettes were lagging behind other makes in TT development, and if you compare their racing efforts with Norton and Rudge this is true. But it was a question of policy. They were still a small firm without outside financial support and quite unable to match the resources of the two main combatants. Wisely, I think, they concentrated on producing road machines which sold well all over the world because of their reputation for quality and speed, and racing machines which would give the private owner a sporting

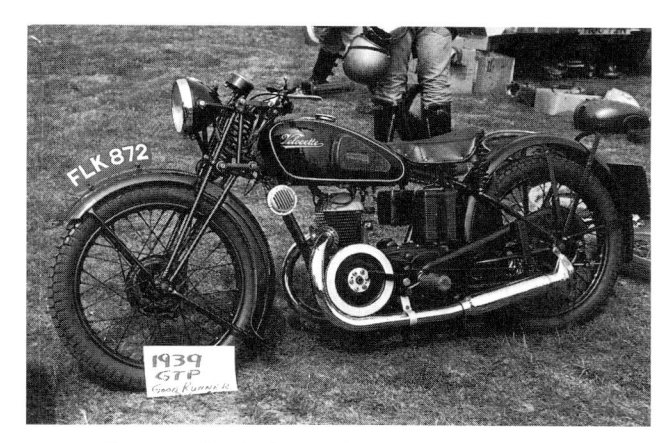

Seen at an Autojumble; a tidy and original looking GTP

chance in any company. And at this period of industrial depression there was a need for a cheaper model than the ohc ones could ever be. Much of their design and development programme had been diverted to this problem. At first it was thought the answer would be in an under 224 lb (low tax) side-valve. Many other manufacturers thought the same and there was a short-lived spate of little side-valves in the early 30s. Phil Irving was given the job of designing what was to be a low-cost, low-performance side-valve, but when an engine was built and installed for convenience in a GTP frame it turned out, he has written, to be a no-performance model. So another engine was designed, this time with push-rod ohv and installed in a new cradle frame which had been intended for the side-valve. This 250 cc engine was to start a completely new line, the M line, and was to be one of the best loved and, at its passing, most lamented Velocettes of all time.

FATHER'S VELOCETTE

Father bought a motor bike,
A "Velocette" by name,
And on that self-same evening,
Learnt to ride the same.

Father was proud as Punch,
Astride his new machine:
Started touring round the world,
Thought it was a dream.

Mother, she had many a fear,
When Father started off,
Tickled the carburettor,
And swung himself aloft.

She thought he'd surely tumble
When going at that speed
Then in a hedge go spinning,
And kill himself indeed.

The children answered, "Never !
On that we'll take a bet:
Father's as safe as we are,
He's on a 'Velocette."

(A few verses on the Velocette, taken from the Veloce booklet 'The World's appreciation of the Velocete, published c.1922)

"Generally Tight Piston"

MANY OLDER enthusiasts, and I am included, mourned the killing off of the GTP two-stroke (the last batch going to Australia apparently fitted with magnetos but that may have been because coil equipment was not available). Mention of the model brought back memories to many of us, often enhanced by nostalgia, with time drawing a merciful veil over the down side. I fell in love with a 1934 model in my late teens, although I had not been impressed with earlier three-speed models. They, I recall, were rough in feel and severely handicapped by the limitations of the wide-ratio gears and the hand-change. The Willis foot-change was always an optional extra but times were hard in the early 30s and few could afford it. Nor was it much help with a gear change perforce so slow.

But the four-speed foot-change, like that on the MOV, transformed the model and I think that the revised porting to cool the underside of the piston might have had something to do with the increased smoothness and "eagerness" of the late engine. It steered like a Velo, and even better after I had replaced the rather heavy front mudguard with a more sporty narrow one (mud on my waders didn't matter in those days but image did), and hummed along at about 45-50 mph up and down dale while you made music with that lovely four-speed box. The original silencers were not the best feature: a clever design which enabled you to completely dismantle them by undoing the clips which held them to the exhaust pipe and held the fishtails in position, but they leaked oily goo at the joints and tended to work loose in use. A pair of tubular silencers from Halfords (our first port of call in those days) made a sharper, more aggressive exhaust note and did not, as might now be expected, affect the performance.

Though second-hand it never gave me trouble with the electrics or the clutch. The dynamo was sometimes a little reluctant to start charging but a burst of revs or a tap on the end cover would always wake it up, and not understanding how the clutch worked I left it alone which is probably why I had no trouble.

Yet Harold Willis, when asked what GTP stood for quoth wryly "Generally tight piston", and when, a great many years later, I asked Eugene Goodman why they dropped the GTP he replied that it had given them a lot of after sale service trouble and the MOV was a much better bike. He left me no doubt that they were, towards the end of the thirties, glad to get rid of it. I was rather shocked at this revelation but when I look back I realise that the market for single cylinder two-strokes was declining, as four-strokes, ohv in particular, forged ahead in design. The noisy, troublesome valve gear which had put many of us off was now enclosed, properly

Final development of the two-stroke. The GTP - virtually unchanged from 1937 to 1939

lubricated, and out of sight and mind.

I was a typical buyer and I moved on to a 250 cc ohv AJS which was the cheapest in the range and it withstood 20,000 miles of hard use and frequent thrashing and then merely needed valve springs and piston rings to regain its original performance.

Thrashing — yes that was what the GTP and, to be fair, all two-strokes of the period, could not take, led to their decline until new technology brought them back with a flourish.

What do I mean by thrashing? Expecting too much from a machine of modest performance, which had charm rather than stamina.

Let me tell you what that long suffering GTP had to put up with. A teenage friend and I were offered an adventure holiday in Skye and my GTP was our only means of transport. Two-up and laden with camping gear we left London at dawn with the optimism of youth. I knew of course that too much throttle too long would cause it to nip up. Coasting with the clutch out and throttle shut would cool it sufficiently for it to recover before you came to a standstill. With long experience of two-strokes I could usually feel a seizure coming on and ease back in time but the excess load meant over driving nearly all the time and the partial seizures became an irritation. Turning up the oil supply of course doesn't help with a GTP. It's not shortage of oil but excess heat that's the problem. Too much oil reduces power and can make things worse. The standard plug looked as if it had been in a furnace, which in a way it had, so we bought a harder grade, a KLG 583, the sort of plug racers used for warming up. Somewhere near Doncaster on the old A1 it seized solid. So solid it wouldn't turn when we dropped the clutch back in. You could just turn the flywheel but it was very stiff.

We took off the cylinder head and knocked the piston down with a piece of wood and a brick and prised off the cylinder barrel. The lofty crown of the deflector piston had caved in a bit with the heat and this had swelled out the top land, so no way would it go up and down. We borrowed a rather rusty file from a cottage and filed the piston roughly round until it would fit easily in the barrel. It took a long time but it was a nice day now the sun had come up, and there was not much traffic in those days. The rings were partly stuck in the grooves with smears of alloy from the piston but we freed them with careful scraping with a pen knife. Believe it or not, the GTP ran and took us on to Skye and back, albeit at much reduced speed. Back in London I bought a pattern piston instead of the genuine Velo one and it had a distinctive rattle which was a good thing really because the rattle increased when you were nearing the seizure point. A death rattle, you might say. It didn't seize but then again I didn't thrash it. It hadn't the performance to which I was now accustomed and I wondered how I could possibly have been enraptured by one.

Piston seizures were still troubling the Service Dept. when Phil Irving went back to Hall Green in 1937 and, after examining a few exhibits, he designed a fixture by which the piston (they made the GTP piston 'in house') could be diamond to a shape which in his words "would compensate for the distortion created by high running temperatures". The shape was finalised when none of the testers could manage to seize a GTP.

I wish I had one of those pistons.

Willis was right, Eugene Goodman was right and Velocettes were right when they dropped the GTP. It had been quite the nicest two-stroke machine of its era but four-strokes had caught up and left it way behind so the only customers were two-stroke fans, largely elderly and content with modest performance. An ideal machine for today's vintage and classic scene, for discovering the delights of the country lanes far from the madding crowd. Sadly there are not many GTPs about.

KTTs and the new M series
There were significant similarities

In my account of the "dog kennel" cam-box Velocettes introduced for the 1933 TT, I mentioned that the cylinder head and barrel had an unfamiliar appearance, unfamiliar, that is, to the student of the ohc Velocette engines that had gone before — and bore a likeness to the M series of pushrod engines under development. Under development, but not as yet announced to the public. In the layout of the finning on the head, the angle of the ports, the inlet port with a pronounced down-draught inclination and the cylinder barrel sandwiched between head and crankcase by through bolts and deeply spigoted into the crankcase mouth, there were all the features which made the first M series engine, the 1933 MOV, such an outstanding design. Features which were still to be found on the last Velocette models. Careful examination of photographs will show that the crankcase was, apart from the extended crankcase mouth, very similar to those of previous models save for alterations to the lubrication system indicated by some not exactly handsome external plumbing. It had long been felt that the original system of forcing oil up to the cam box via the vertical drive tunnel, bottom and top bevels, plus the shaft being submerged in oil at around 12 lb per sq in, though excellent from a reliability point of view, did result in a power loss. I well remember the importance attached to getting engines not just warmed up but stinking hot by short-circuit stars in the early 30s.

I remember, too, that private owners had carried out modifications to by-pass the oil to the cam box. J.M. "Spug" Muir, one of the leading private owners, was I believe a pioneer in this field. Now with this new works design the first real factory change in the lubrication system could be seen. Oil for the cam box was piped up from the pump to a quill in the camshaft and any surplus in the upper bevel box drawn off by another external pipe. The vertical-drive shaft was no longer under pressure and received adequate lubrication by spillage at the upper bevel box.

Still intrigued by the external resemblance of the upper works of these engines to the M series, I wondered if there had been a policy reason behind it. The desire to make a racer look a little like the production models which were to be introduced? It was an unworthy thought, a thought I should never have entertained in the case of a firm like

Enduring memorial to the genius of Percy Goodman: the ageless, tough vintage KTT engine

Velocette. Charles Udall, who did much of the drawing-board work on these engines, made it perfectly clear when I quizzed him. From the tone of his voice it was immediately apparent that Velocette never did act from such base commercial motives (they might perhaps have made more money at times if they had). No, the reason, as he explained, for the similarity was due to two factors. One was that casting techniques had reached the point when that was a good way to make a cylinder head and any superficial resemblance in styling was due to the same designer being on the drawing board... the artist's signature, in fact.

The head was cast in alloy-aluminium bronze as on the Mk. 4 KTT, which was the production racer. But it was the cam box which was the really interesting feature for a serious attempt had been made to overcome the oil-slinging characteristics of all previous designs, Velocettes and others. Where the rockers protruded from the cam box it was impossible to seal them completely or for long. Velocettes extended the cam box to completely enclose the rockers, and operated the valves by means of short tappets, the hairpin valve springs remaining exposed as at that time in the development of valve springs it was considered essential to expose them to the cooling breeze. The basic design was a natural for the future development of a dohc engine, which Velocettes did before long, only to drop it like a hot brick after the "first-off" disgraced itself. Nortons then took up the idea and with the painstaking development of Joe Craig turned it into a world-beater.

Full enclosure of the rockers posed a tappet problem which Velocettes solved neatly by the use of an eccentric rocker spindle, which proved so successful that it was incorporated in all their subsequent ohc designs. Of necessity, the cam box overhung the head fore and aft — the gable end of the "dog kennel" — and was supported at either end by short links to the head. This disposed of any problems of differential expansion between head and cam box.

Because the new engine was tall, the Mk 4 type frame had to be modified. The tank rail was discarded, the top tube made heavier. A heavy single front down tube replaced the traditional twin side-by-side tubes. It was not yet a full cradle frame but the lower torque stays ran to the front of the crankcase and engine plates brought well up in front produced virtually the same effect. It was, in fact, an interim frame — the rear half of the old frame and the front of the new cradle frame which was being developed for the roadsters. The gearbox was as before, with the external Willis drum positive-stop mechanism on top.

A really novel feature was the fitting of a rev counter in the tank top with a drive from the cam box. It was stressed that the instruments would only be used for practice. The space in the tank would be needed for fuel in the race.

Altogether, the works models for the 1933 Junior showed a new line of thought and presented a new image. The writing on the wall was clearly that the days of the old K type vintage engine, changed only in detail since its inception in 1925, were numbered. Willis, never afraid to poke fun at the firm's products, is reported to have referred to the dog kennel engines as "Velosacoche" or "FNocette" because of a superficial resemblance to Motosacoche and FN engines suggested by the lanky looking barrel and overhanging cam box.

Velocettes didn't pull it off in 1933: it would have needed a super bike and a super star to take on Nortons in the ascendancy with riders like Stanley Woods, Jimmy Guthrie, Jimmy Simpson and Tim Hunt.

They did the next best thing, by taking the team prize and filling the bulk of the places below Guthrie's third ... fourth, fifth, sixth, seventh, eighth, ninth, 10th, 13th, 14th, 15th and 16th. Wal Handley at first led the Velocette pack which harried the Nortons, followed closely by Alec Mitchell, a clubman rider who had won a work's ride by brilliance in the past. Tyrell Smith followed, a little slower. This position was maintained until the fifth lap when Mitchell got in front of Handley to lay fifth behind Jimmy Guthrie. Handley slowed mysteriously on the next lap, and Tyrell Smith and Gilbert Emery moved ahead of him. The last-lap failure of Jimmy Simpson's engine lifted Mitchell to a richly deserved fourth place. Stanley Woods had led from the start with Hunt and Guthrie second and third except when Jimmy Simpson intervened. It turned out that Handley had come off at Governor's Bridge on the fourth lap and had spent a long time at his pit when refuelling. Perhaps a little straightening-out operation.

The "dog-kennel" engine returned to the Island in 1934, this time with a big brother, 81 x 96mm, 495 cc edition. This bored and stroked version looked identical to the three-fifty, and in fact the bicycle parts were the same.

Both engines went into a new full cradle frame which was virtually the one which was to become standard for the 500 cc pushrod MSS and a redesigned ohc roadster, the Mk. II KSS. It had a heavy single down tube and top tube and the seat post was now vertical behind the gearbox. Engine and gearbox were retained by plates which encircled the gearbox and formed a semi-unit design. Velocettes scored another design "first" with a then revolutionary rear hub of conical shape, the cone being cast in elektron and the steel brake drum and sprocket bolted on.

There was another "first" too in the combined saddle and mudguard pad. "Saddle-cum-mudguard pad", the reporters called it, rather clumsily. "Loch Ness Monster" Willis called it. It was not taken very seriously from a racing point of view but a caption writer in *The Motor Cycle* showed unusual vision by writing "The arrangement might well make an appeal to pillion-carrying owners of touring machines". This prophecy passed unnoticed by most people.

Velocettes soon gave up the idea and when they next redesigned the racers they dropped it because, it was said, it was not in keeping with their production machines. It took an accessory manufacturer, one J.R. Ferriday, to realize its real worth, christen it a "dualseat" and make it standard equipment the world over. Velocettes, it is said, sold the patent for a song.

The five-hundred was only just completed in time. In fact, it was said that the engine had not been tried on the road before it reached the Island. But for all that Walter Rusk was to finish third and to go on and win in Ulster. It emerged that apart from the bigger bore and stroke and a certain amount of strengthening of the bottom end by means of heavier shafts, and a double row roller drive-side main bearing, the Junior and Senior engines were very similar. The cam box was, in fact, interchangeable. A novelty on the five-hundred which was to start a minor fashion was the use of a BTH magneto firing two spark plugs simultaneously.

By stretching a basically three-fifty design to five-hundred size and installing it in the three-fifty frame, Velocettes of course achieved an excellent power-weight ratio. The Junior Nortons, by comparison, were more like scaled-down five-hundreds.

In practice Gilbert Emery, who used to have a motorcycle business at Colwyn Bay, topped the first Junior times.

Walter Rusk, a newcomer to the Velocette team and tipped as "a second Stanley Woods", was next best but their performance did not occasion much surprise because the Junior Nortons were not out. When on the next Junior outing Emery did it again, this time beating Jimmy Guthrie, and Les Archer was third, Velocette hopes looked rosy. Emery, however, overdid it after making second ftd on the five-hundred Velo, and came off, hurting himself enough to put him out of the race. Walter Rusk had been first to show how the five-hundred could go, making second-fastest Senior practice time in 28 minutes 29 seconds, comparing with Guthrie's 27 minutes 1 second, and Alec Mitchell on another five-hundred Velo was fourth-fastest on his first clear run. Lest it should be thought that the 1934 practising was a simple Norton-Velocette duel, it should be pointed out that there was a real dark horse among them in the stocky shape of Ernie Nott on a brace of Husqvarna vee-twins, and he really upset the tipsters by topping the Senior list on the third day of practice. So when Junior day dawned the feeling was that although a Norton would no doubt win — Norton supremacy was now so firmly established that it was a pardonable assumption — it was not at all certain which make would follow them home. And as it turned out it was Nott, fighting a gallant lone hand on the Husqvarna, who came third behind Guthrie and Simpson. Velocettes brought up the rear, filling most of the remaining places in much the same way they had done the year before, taking the manufacturer's and club team prizes.

Les Archer at first led the Velocette "hounds" baying at the heels of the leading Norton trio of Guthrie, Simpson and Handley, and he held fourth place for three laps till ousted by Nott. Harold Newman, who first rode in the TT the year before and finished eighth, on a Velocette, then showed brilliant form by getting past Nott. Wal Handley fell off and for a lap Newman lay third but on the last lap Nott just pipped him by 14 seconds. Alec Mitchell was fifth, Les Archer sixth, Walter Rusk seventh. After that it was the usual Velocette benefit interrupted only by George Rowley in 11th spot, on an AJS, of course, and Charlie Manders in 14th on a New Imperial. A Norton was last this time, H. Pilling being the unfortunate. Most of the place men were on over-the-counter Mk. IV KTT models.

The Senior was much more exciting, for Stanley Woods was on a big Husqvarna and chased the winner, Jimmy Guthrie, relentlessly — though perhaps hopelessly — for six laps before running out of petrol. Nott was not on his big Husqvarna very long; the gearbox failed before Kirkmichael. The big Velocettes were slow to come to the fore, rather as the early vintage ones had been tardy, and their leader, Les Archer, who had taken over Emery's work's mount in place of his own three-fifty, was 11th on the first lap, while Arthur Tyler on a Vincent HRD JAP was way ahead in fourth place, chasing Jimmy Simpson.

By the third lap the surprisingly fast JAP engines in various frames had blown up and the Velocettes moved up into their accustomed position. Rusk took over from Archer and both of them displaced Vic Brittain, better remembered now as a trials man but then a Norton teamster, and so when Stanley retired on the last lap Rusk became third and Archer fourth, and it should be noted that Newman on a three-fifty Velo finished eighth while Alec Mitchell on the remaining five-hundred was 10th. Rusk's performance was the more remarkable because an oil return pipe came adrift and plastered him and his machine with oil.

It was noted afterwards by a diligent scribe that only one Velocette out of the 17 running in the two races — many of them privately owned; nay, most of them privately owned — retired with mechanical trouble and that was due to the loss of a nut in the gear-selector mechanism. It speaks

Sports-touring KN model differed little from the KTT, apart from tyre and wheel sizes, and rear carrier on the roadster

volumes for the mechanical excellence and the careful preparation of racing machines under the direction of Harold Willis.

The ultra-fast Ulster GP course again gave Velocettes an opportunity to get their own back on Nortons. Walter Rusk won the Senior GP at 88.38 mph and put in the first lap at over 90 mph, while Newman was second in the Junior class, a mere three seconds behind the winner. The real significance of this season's racing — defeat in the Island and victory in Ulster — still did not become apparent to Velocettes. They were as ever looking for more speed and overlooking the fact that the Island course puts an even higher premium on cornering power.

At this time a really-up-to-date roadster ohc machine was being designed finally to replace the vintage type K models. This machine, the Mk. II series of KSS and KTS (identical machines apart from mudguarding and tyre sizes), retained the same model symbols as the vintage ones but differed in character. The vintage ohc models were essentially super-sporting roadsters intended for the "have-a-go-at-everything" clubman. They were refined enough — more so than most other makes — for ride-to-work and holiday touring yet were snappy enough and cobby enough to double as short-circuit racers, grass track and scrambles mounts and trials mounts in the winter. First-rate all-rounders. Admittedly, they barked a little, hopped a little with their short wheelbase, and slung a little oil, but these things did not matter to the ardent clubman.

But by the middle 30s the scene had changed. Competitions were already becoming a little more specialized. The day of the all-round clubman with one machine was passing. Clubmen tended to specialize in one type of event. Tended, in consequence, to settle on one specialized machine. Moreover, there was growing demand, fostered continually by the weeklies, for quieter, more refined roadsters for the growing body of motorcyclists who did not want to compete but liked to tour and to go and watch events. So in redesigning the ohc model Velocettes set out to provide a luxury high-speed roadster rather than a super sports roadster. And in this they succeeded most handsomely although many enthusiasts, accustomed to a succession of vintage type cammy models, were perhaps disappointed when they found the new model had gained weight, bulk and sophistication and had lost its sporting image.

At this stage in the saga it is fitting that a new character should be mentioned in more detail than in previous pages. Tall, studious, retiring Charles Udall was to play an increasing part in Velocette fortunes as the next decade came in. Charles Udall had been around quite a long time, as a matter of fact, but always behind the scenes. Like most of the Velocette key men he had come to them because he was a motorcycle enthusiast... not a competitive one, for he never rode in sporting events, but the more academic type who liked riding but was absorbed in the technical problems of motorcycle design. He started in the repair shop, which was the only vacancy available in 1928, and this was a good

Carlisle's Billy Tiffer Jnr. was one of life's all-round motorcyclists. Photograph at left shows Billy Jnr starting in his first trial on a model U, Good Friday 1928. In 1936, he became the only Velocette rider to win the Scottish Six Days Trial. He took over his father's entry — and his KSS trials iron (above) — when Dad fell ill with food poisoning and went on to win with a loss of four marks, also winning the Scott trial in the same year

thing for it gave him a practical approach to design.

"In the repair shop you get to know what motorcycles are really made of and if you are going to be a designer it is good training. You learn, for instance, not to put nuts in places where you cannot reach them. No designer can design a good motorcycle unless he can ride one."

An opinion to which I, for one, subscribe.

Although he had been trained as an engineer he had to wait until an opportunity presented itself for a better job.

"One day I heard that there was a vacancy in the drawing office. I asked the foreman about it but he was not helpful. But I had been repairing the brakes on Mr. Percy's Rhode car and I tackled him about the job. 'Can you draw?' asked Percy Goodman. 'Yes' I replied. 'Then start on Monday' and that was that."

I doubt whether Percy Goodman dreamed of the importance of his snap decision.

I do not think it is right to conjure up the picture that "drawing office" usually suggests. Certainly not in terms of spacious, well-lit offices with lines of draughtsman bent over drawing boards. It was not like that at all and for many years Udall not only had to do the drawings but make the blueprint copies as well. At first it was a matter of detailing some of the ideas of Percy Goodman, Eugene Goodman and Harold Willis, and in this way he was connected with the various gearbox experiments, but his first real chance came when the two-fifty ohv was designed. He liked the high-camshaft design mooted by Eugene Goodman because he had been impressed with the design of the Riley car which Willis had owned, a pushrod design with two camshafts mounted high up in the block and a hemispherical head.

It was a good decision. One that changed the fortunes of the company and finally succeeded beyond all expectations in its post-war guise as the Venom and Viper. I attribute the success of the M range from 1933 onwards to two features of design above all others. First, the excellence of the timing gears in terms of silence and long life and secondly the equally indestructible qualities of the overhead rockers. There were, in the 30s, several high-camshaft designs, no doubt inspired by the Velocette, but none of them were very good, long lasting or silent and they have gone without mourners.

The unique feature of the M range timing gear was the use of helical gears of fine pitch, well supported independently of the outer cover of the crankcase wall on one side and a stout steel plate on the other side. The first MOV models to be produced in 1933 actually had spur gears but these were soon changed for helical gears. I wondered why?

"The prototypes ran quietly with spur gears but when we got into production we had trouble with noisy gears. We

Tiffen the racer. Billy Tiffen Jnr with one of the Mk. 6 KTTs in the Island, during the 1936 TT

found that the prototype had a thicker crankcase wall due to a pattern error and the production ones of the correct thickness acted more like a sounding board."

I cannot imagine any other firm going to the expense of substituting expensive helical gears to cut timing gear noise on a low-cost machine. Thicker castings perhaps.

The ohv rockers were unique with their light, large-diameter centres and light arms. The bearing area was unusually large at a time when bushed rockers working on fixed spindles of perhaps ½in were considered adequate. The result was that they were quiet and, enjoying pressure lubrication, were virtually everlasting. When I asked Udall about his feature of the design he did not go into detail.

"Rocker design is a case where some people will not take notice of the obvious."

He is not the kind of man who will waste words explaining what should be obvious. His approach to detail design was undoubtedly influenced by his mentors, Percy Goodman, Harold Willis and Eugene Goodman. A philosophy which sought to strike to the root of a problem and disregard the conventional approach. "We did not believe anything we were told and only half of what we saw", he said.

Thus it was over the matter of the hairpin valve springs.

When they started using hairpin valve springs they were troubled with breakage and rapid wear where the ends fitted into the retaining blocks. The Guzzi was always cursed with spring breakage. It was regarded as one of those things. Udall studied the action of a hairpin spring from a geometrical standpoint and found that by having the correct size and number of coils the wear and tear could be avoided. Even designed a set of springs for Fergus Anderson's Guzzi. They did not break.

The MOV Velocette of 1933 was a complete success mechanically, but two-fifties were still frowned on as being miniatures. How about a three-fifty one the same lines? Udall remembered the Willis long-stroke of 68 x 96 mm. It worked all right. There was no snag about the long stroke. So an MOV was stretched by increasing the stroke, lengthening the barrel and conrod — and the MAC was born. Everything else was the same. The result was a charming machine with light weight, bags of bottom-end power and the characteristic M type silence and longevity.

I wondered if the three-fifty version had been in mind when the two-fifty was designed, but apparently not. Like many good things in the motorcycle world, the MAC was born out of expediency rather than design. No one, of course, ever thought of racing the pushrod models... It just wasn't done, except in Australia where the outlook was less inhibited and a tuner got an MOV doing over 100 mph, a claim which Hall Green frankly disbelieved at the time. Pre-war, we enthusiasts hardly gave the pushrod models a second glance competitionwise. Status demanded a cammy or nothing. How wrong we were. After the war when sporting machinery was at a premium a few desperate souls souped up MOV and MAC models for grass and scrambling and found them very sprightly.

After the commercial success of the small pushrod models and the vindication of the high-camshaft engine and its lightweight valve gear, a five-hundred version was an obvious step to widen the range. This time the basic design was retained but enlarged throughout. It was not seen as a high-performance sports machine but as a refined medium-performance model for solo and sidecar use.

"I wanted a sweet, smooth, flexible machine. If I liked it I felt sure others would like it," says Udall. Note here the reasoning of the rider-designer, the basis of Veloce psychology.

"The first experimental one was rather inflexible. An engine-shaft shock absorber helped, but it was still necessary to manipulate the ignition intelligently. Automatic ignition control was universal on cars and I was always interested in car design... aircraft design, as well. I went to see Mr. Griffiths of BTH and they designed the first auto ignition control for a motorcycle."

The result satisfied Udall's personal desire for a smooth,

flexible mount and pleased countless motorcyclists who liked the new experience of the docility of a low-compression side-valve allied with the top-end performance of a contemporary ohv.

Quote from a road test: "At idling speeds the engine is extremely smooth, mechanically silent and docile. It gives the impression of plenty of power at low speed engine revs and its response to the throttle is surprisingly good.

"Such is the flexibility and steam-like power surge of the engine that gear changing can be forgotten from any speed above 16 mph in top gear. On the road the extremely docile engine can be pushed up to a maximum which completely belies its side-valve characteristics lower down the scale.

"The Velocette can be ridden comfortably up to 70 mph in third gear before changing up and in this gear it had the performance of a highly tuned sports machine."

After all these years it is difficult and invidious to be specific about "who did what" in regard to the various designs which emerged from this closely-knit firm where all strived for the common goal of producing better roadsters and more successful racers. An "official" history commissioned by Veloce Ltd. for the firm's jubilee in 1954 states:

"For the 1933 season a two-fifty model known as the MOV was added to the range and four-speed gearboxes standardized on the ohc range. The MOV was designed by E.F. Goodman. The first of the M series of machines to be mentioned later was a single-cylinder 68 x 68mm (250 cc) and the machine was new from start to finish. It was the first engine to have oil pump lubricated rockers and valve guides. It is interesting to note that Mr. Charles Udall, who is now (1954) Veloce Development Engineer, assisted in his youthful days with the drawings of this model. It was also the first production model to be fitted with a propstand and positive-stop foot-change. The gearchange was a four-speed type and was bottom mounted in the cradle frame."

"Becoming popular at once, the MOV was followed in 1934 by the second M-type machine, the MAC. This machine was the same as the MOV except for the fact that it was of 350 cc achieved by lengthening the stroke to 96 mm."

"The third type M engine, the 500 cc MSS, was of the same general construction but was of 81 x 96 mm. bore and stroke and the frame, wheels and gearbox were identical to the corresponding parts used in the KTS model 350 cc ohc produced from 1936."

"The MSS was first sold in 1935. Outstanding feature was that H.J. Willis evolved the idea of having the ignition advance and retard automatically controlled and this was carried out by BTH in conjunction with Veloce Ltd. Exclusive rights were given up during the Hitler war for the national benefit."

Interviewing Eugene Goodman
A down to earth artist in metal

Father and son. A wartime picture of Eugene and Peter Goodman

At the beginning of this story of the Velocette *marque* I stressed that, more than a history of a particular make of motorcycle, it is the story of the Goodman family, dedicated through three generations to the task of building motorcycles. Motorcycle of high quality and great individuality and that subtle character possessed only by machines, which are built by practical motorcyclists. Percy Goodman we have come to know, as the story has unfolded, as the designing genius behind the immortal "cammy", an indefatigable defender of British prestige in the racing field between the wars. But his younger brother, Eugene, had been but a shadowy figure in the background since the early 20s when on the little two-strokes he was a successful competitor in long-distance trials. True, I have related how at the end of the vintage period he designed the high-camshaft MOV as a second string to the ohc stable and how this basic design, developed into 350 and 500 cc sizes, eventually outlived all his brother's designs, but until I had the pleasure of meeting him I was unable to get his part in the Velocette saga into perspective. Nor, I must confess, had I been able to build an accurate mind's eye impression of the man. It must have been the christian name Eugene which led me to imagine an artistic type. Artist in a way he was, but with metal and machine tools. His studio was the grimy factory floor, and background music the hum of machinery and the thump of presses. A tough, bluff little man calling a spade a spade in an unmistakable "Brumma-gem" accent. A man who in his retirement looked back on both triumph and disappointment, as do many other men who have devoted their life to the motorcycle industry with its see-saw of boom and slump.

Time was short and I had to come straight to the point in my questioning.

"How did you come to design the pushrod engine... there had not been anything like it before although it was soon copied, though with little success?" I asked.

"I was not satisfied with the amount of business we were doing. My brother Percy and Harold were wrapped up in racing and my job in the beginning was to make the bits for them. But there were men working in the factory who took

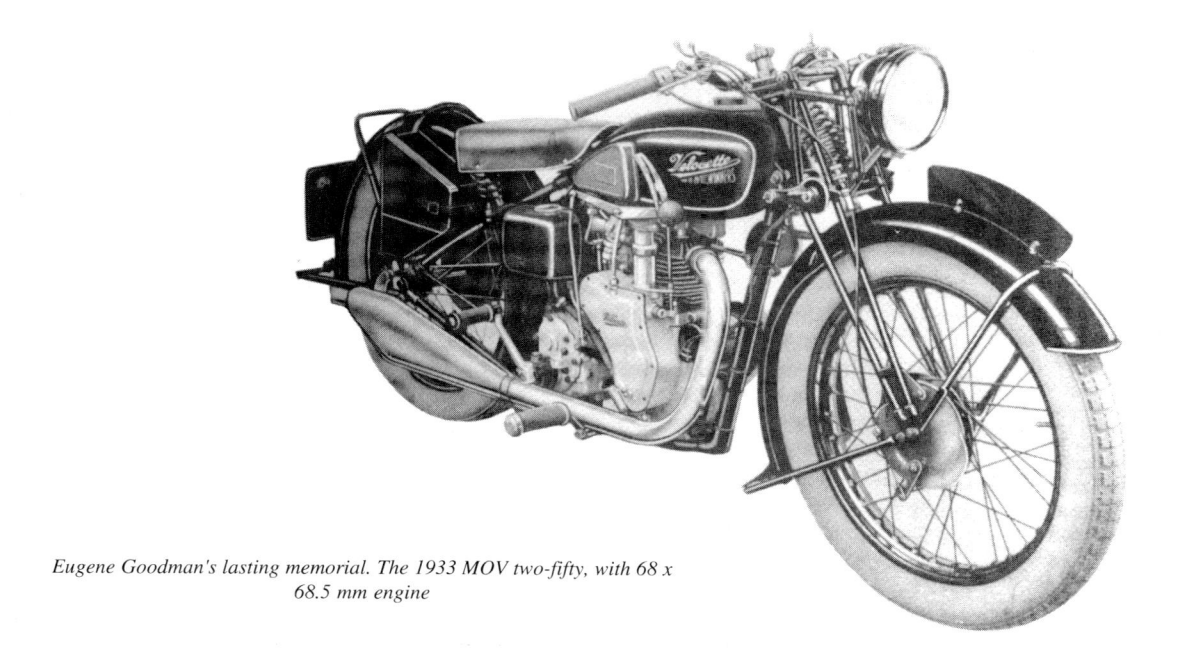

Eugene Goodman's lasting memorial. The 1933 MOV two-fifty, with 68 x 68.5 mm engine

home more money than me. I wanted a model which would sell in bigger numbers than the overhead-camshaft ones I had to fight to get my way, even threaten to leave. I could have easily got a job somewhere else."

"Did you get the idea of a high-camshaft engine from the Riley car?" "I've no idea what timing gear the Riley engine had. I studied patents relating to engine design... hundreds of them,. I had a book full. I came to the conclusion that if I raised the timing gear with intermediate gears I could get nearly the same effect as an overhead camshaft."

"And your original design for the timing cover is still being used today. Even the timing cover of your pre-war MOV will fit a Thruxton model."

"Yes, I suppose it will. We used to call that timing cover the 'Map of Africa'."

"Before the war you had a prototype machine running about with a pressed-steel rear end and pivoted-fork rear suspension. Why were you experimenting with this when you had rear suspension on the Mk 8 which was years ahead of its time?"

"It was my idea to save cost in production. I wanted to get away from the tubular frame with all its bits and pieces. The rear end was made out of two pressings welded together. We used the same idea on the LE. Irving designed the adjustable spring frame."

"I note that the experimental machine had a two level dualseat which fitted the pressed steel frame. Was this a development of the racing seat... the Loch Ness Monster?" "We sold the patent for the original seat. This one was different but we didn't patent it."

A full description of it was written by Graham Walker, then Editor of *Motor Cycling*, in August 1942. It was quite a "scoop" because no mention of it had been made in the motorcycle papers before. Quite a few people must have seen in on the road in the 1938-9 period because, it was subjected to an exhaustive road-test programme in the hands of Franz Binder who raced Velocettes in the Island and on the Continental Circus. But in those days motorcycle pressmen were oh so discreet. They never published anything about a manufacturer's plans without permission, Graham Walker got his scoop like a real live newshawk. Remembering seeing Binder on the machine in the Island he searched the patent files and found three patents relating to it in the joint names of Veloce Ltd. and P.E. Irving... Irving of course worked on Velocettes before joining Philip Vincent to assist with the design of the Rapide. Armed with the facts, he approached Velocettes and got permission to write his story. It took a lot to stop Graham Walker when he was on the scent! The first patent concerned means of adjustment for varying loads. It was the theoretically excellent scheme of moving the upper abutment of the suspension legs in an arc to vary the mechanical advantage. Altered only in constructional details on the LE and the singles. It is an infinitely better way of catering for different springing requirements than altering the preload or rate of springs in telescopic units.

Eugene Goodman gives the credit for this idea to Irving.

The second patent protected the method of rear-chain adjustment. Instead of adjusters in the fork ends, a throw back to pedal cycle practice still used by manufacturers who I think should know better, the swinging fork pivot was mounted eccentrically so that in no circumstances could wheel alignment be lost.

Why Velocettes dropped this eminently satisfactory method I do not know but I think I am correct in saying that it was next adopted on the post-war DMW lightweights, then later it was used by the Rickmans for their Metisse frames.

The final patent covered the pressed-steel stressed-skin rear construction. The patent drawings show it with a saddle and pillion pad but the prototype machine had a foam rubber-moulded dualseat very like the two-level seat fitted to post-war Velocettes for a period.

Not included in the Veloce Ltd. patents, though perhaps covered by the makers. Messrs. Ferodo, were the suspension units. They looked like conventional post-war units but were quite different in the damping mechanism. As one might expect from a firm specializing in friction material they were friction damped, not oil damped. The damper "rod" was guided by two Ferodo sleeves expanded by an internal clock type spring. It was simple two-way friction damping with the same characteristics as a scissor friction damper but it was neat and well protected from the elements. No lubrication was necessary, of course, as the only contact between the telescopic members was via the Ferodo material. I suppose Veloce Ltd. were wise to drop the pressed-steel rear frame idea for roadsters. The motorcycle enthusiast constantly pleads for originality in design but will only accept it when it has become commonplace. Restyled, it might have been viable in the short-lived "bath tub" enclosure fashion of the middle 50s but would be "square" again today.

I have come to believe that Velocettes under the aegis of the Goodman brothers Eugene and Percy, loved designing motorcycles more for designing's sake than for the rewards, be they financial success, racing victory or public acclaim. Aided and abetted by Harold Willis, Charles Udall and for a period Phil Irving, they must have sought the satisfaction

The MAC was more than a "cooking" motor

WRITING as a once-typical pre-war clubman looking for a ride-to-work and have a go at anything weekend bike, be it trials, grass track racing or courting, I very much regret that I never showed any interest in the push rod models. I had tried a Mk. II KSS and been disappointed to find it an overweight tourer and no longer a sports bike in the manner of its wonderfully versatile and cobby cammy forebear. I am afraid we club enthusiasts never even considered a MAC as a substitute. Had we known then how easy it would be to convert the rather "cooking" MAC to a sports bike capable of seeing off a new KSS we might have stayed faithful to the marque. But Hall Green did not present it as a sports bike. They could have made a MAC SS in addition to the standard model but they didn't. Perhaps they feared it would ruin the sales prospects of the cammy job. Nortons pursued the same policy with their ES2, only selling it in low performance guise after 1928 so as not to compete with their camshaft models even when those only sold in penny numbers and were loss makers in the long term. I did not discover the MAC potential until looking for something cheap for my teenage son to sprint. I was given the remains of a 1935 MAC. A friend with a big old-fashioned lathe turned ¼in off the conveniently round cylinder base to up the compression and with a quite Victorian pillar drill, bored the inlet port out to 1in and a sixteenth to take a bigger Type 6 carb. I reprofiled the inlet valve with a file while spinning it in my pillar drill, slipped an extra spring inside the old and frail valve springs. The resulting urge combined with the light weight after we had discarded the unessentials was most encouraging and too much for the clutch, of course. Mike Tomkinson, an old friend, gave me a KSS clutch and wheelie take offs were easy. A conversion to methanol by leaving out the main jet and the throttle needle and filing the slide cutaway to something like 6/12 made it a real flyer.

Rebuilt in vintage racing trim, it went on to become a very competitive post-vintage mount. If only we had known pre-war what potential the MAC had just waiting to be released.

the artist feels in creative work rather than the commercial kudos and the flesh pots which are more likely to attend the mass production of conventional machines built down to a price by costing experts and lauded to the skies by publicists.

To support this argument I would recall the Viceroy scooter... now virtually forgotten and completely wasted in its day on a market which was technically dumb and sometimes so daft that it ended up trimming its machines with fur, sticking up dummy aerials and dangling tassels from handlebars!

The Viceroy beneath its rather forbidding exterior was a study in functional design and the only two-wheeler to have been powered by an opposed piston inhaling through diaphragm valves.

And there was the short lived Valiant, a miniature BMW if you like, made for a market which did not really exist and too extravagant in design and cost for the market that did.

Even the LE really missed the commercial boat until it was taken up by the police. Only the Goodmans would have conceived a utility machine of such Rolls-Royce specification that it was like offering an exquisite Swiss watch to a navvy. All these "lost causes", to quote Lord Montagu, must have provided tremendous pleasure though little reward in the end.

The 1947 MSS five-hundred: for touring types

The M Super Sports

IT IS A SHAME Velocettes did not offer a sports version of the MAC (though I can see they did not wish to compete in the market with the KSS) but this consideration can surely not have applied to the MSS which also suffered through being given a touring image. Always it was presented as a gentlemanly, de luxe tourer, sporting only in performance comparable with more sporting machines from other manufacturers. Again looking back to the day when I was a potential clubman customer (or would have been if I had the money), I can assure you that neither I, or my friends put the MSS on our shopping list. Not even when they were second hand though there were few used ones about, for few new ones were sold before the Speed Twin came out and became a market leader. But if the MSS... what a misnomer the SS bit was... had been offered with a 21in front wheel with a small section rib tyre, narrow mudguards, all from the KSS and a mildly tweaked engine with higher comp piston, bigger carb and a more sporty cam from the wide range in the stores, we might have considered it! Oh, and a manual magneto. The auto advance gave wonderful flexibility but we young bloods preferred to juggle our own spark, even if the auto could have done it better.

They did, as I mention elsewhere, make up such a sportster for young Peter Goodman using apparently one of the Mk. 6 racing frames (very like a KSS or MSS frame) and a cylinder head cast in bronze, but their marketing policy in the run up to the 40s seems to have been to get away from the sporting image and go for the de luxe roadster for the more mature rider.

But there was considerable potential in that long-stroke motor. The best example that comes to mind ,because it also demonstrates the lasting stamina of the engine, was that of Len Collins who dominated the 500 cc (and often beat bigger bikes) class in sprinting for several years after the war. Collins bought his MSS new in 1936, fitted a sports touring sidecar and with mild tuning made ftd in a couple of kilo sprints for clubmen at Brooklands. The outfit was used for transport during the war and when sprints were restarted Collins substituted a plank for the sidecar, put the compression up for methanol fed through a TT carb, and fitted an improved cam.

Without any drastic weight saving (the heavy standard steel tank was always used) it was the machine to beat until supercharged twin specials took over. I know for I could not get anywhere near his times with a 1000 cc racing Brough Superior outfit! The compression ratio was edged up from time to time but the iron head was re-used, now fitted with bigger valves. The original big end and con rod were still in use. The best indication of its capability was an officially recorded speed of 96.4 mph over the kilo finish line at Brighton speed trials. It was then turning over at 7,300 rpm.

When contacted in 1993, Collins was in his 90th year and still had the Velocette outfit as last used "standing in the garage rather forlorn". He wasn't giving away any tuning secrets, merely said getting rid of any unnecessary weight such as mudguards was always a help.

L.W.H. Collins on his pre-war MSS in its final form. Passenger was Len Terry

Eugene would have been surprised... the David Holmes MOV/MAC racers

YES, THERE was a mighty quick MOV in Australia in the later forties and the man who built it and rode it to nearly 50 wins up to 1957 was Les Diener who therefore became a legend in his lifetime. It's usually said that the best specials really look like factory jobs, and from photographs you can certainly say that about this Velocette special. But you ought to add "like an Italian factory special" because Diener built a gear driven 'double knocker' top on the MOV bottom and in its last form with duplex 'Featherbed' type frame it looked more like a Benelli racer than any Velocette ever. The fact that it revved to 9,000 rpm and with 'dustbin' fairing reached 116 mph showed what could be done with using 'state of the art' valve gear. David Holmes' achievement in getting comparable performance with push rods is the more remarkable.

No one that I can remember bothered to tune an MOV or a MAC for racing pre-war, with the possible exception of the Pike brothers although they soon switched to two-valve Rudge machines. There were plenty of "proper" racing two-fifties about then... Manxman, New Imps, radial Rudges and cammy OKs. As for 350s, there were the cammy Velos, Nortons, Ajays and more OKs. We were spoiled. After the war it was a different proposition. There were not enough pukka racing 250s left. So tuners turned to the pushrod models. Most Velo fans cut down cammy 350s but one or two turned to the MOV. Velocette agent Arthur Taylor developed a very quick MOV for Cecil Sandford to ride, and to the consternation of the factory, who of course didn't regard the MOV as anything more than a ride-to-work bike, got it going well enough to challenge the works cut-down Mk. 8s.

I heard about the Dave Holmes MOV from a mutual friend who said it was doing 10,000 rpm. I didn't like to "pooh pooh" this but figured it must have been running light and not producing much power at revs like that. And I figured it wouldn't be long before it went bang. What I didn't know was the engineering capability of Dave Holmes who steadily developed it to a stage when it is mechanically very reliable and devistatingly fast. Over the years he has traded some of the revs for more useable

Giant killer of Historic racing in the 80's when ridden by Stephen Tomes, the Holmes MAC produces more brake horse power than a Mk 8, and with much less weight is probably the fastest 350 Velocette ever on present-day short circuits

power on a circuit. Now it is geared to rev to 8,500 at which it is doing 100 mph. When pushed... after a bad start, for instance... it goes a bit over that, to 105 mph. This is on alcohol... methanol. Before that both bikes, the MOV and a MAC, had alloy heads, go-fast goodies made by an East London specialist dealer named Woods in the 1950s, and because of them the Holmes team, father and son, became involved in a specification dispute with the Vintage Club's Racing Section.

The regulations stating what you can modify and what you can't are a bit complicated and the Holmes men fell into a kind of vintage generation trap. Vintage bike regulations which go back more or less to the beginning of organised vintage racing permit the use of components which are of a type used by the maker originally. The wording and the interpretations are loose, to help people keep the older bikes on the track, and were originally only applicable to pre-1931 bikes. When for racing purposes the vintage deadline moved to 1934, to bring in the bikes of the early 1930s which were being left at home in favour of more advanced late 30s and up to 1948 models, the Holmes Velos could use Velocette parts if "of a type". But an overriding regulation stated that if later major parts are fitted they update the machine. The Woods heads were of course a major component and they therefore updated the bikes to the 1950s. But the post vintage and post war regulations are a bit tighter. They say that major components must be from the same maker as the machine. You can fit later parts, thereby updating the machine, but not parts of another make.

The Woods heads were so much like Velo heads that, painted black and with the bigger fins trimmed to the size of the cast-iron Velo head, they might have got away with it. But when you are out to win and not just have a bit of fun you have to be like Caesar's wife, beyond suspicion. Finish well down the field and no one will give your bike a second look. Win and go on winning and your bike is subject to the closest scrutiny. But with iron heads they could use methanol and stay in the up-to-1948 class.

And go faster.

"The Woods head had a bigger inlet valve, 1.54in against the standard 1.44in valve, and the biggest I can use with the iron head is 1.5in, but the shape of the Woods port is not as good as I can get by modifying the standard port" says Dave.

There is no doubt that the remarkable performance of the 250 in particular, which at the time of writing could pass the best two pre-1948 Triumph 500 twins ... those of Derek Pollard and Tony Wilmott ... down the straight, was mostly due to Dave's understanding of carburation and, to use a fashionable word in the car world, valve "management". Naturally he makes his own cams by building up on a Velo cam to increase lift (7/16in on the 250 and 15/32in on the 350 as compared with the standard 5/16in). His own valves have 9/32in stems and are closed by S & W Gold Star-type springs. Everything is done to reduce reciprocating weight, of course, and though he originally made his own rockers with offset to increase valve lift and to do away with the not very clever adjuster at the valve end of the MOV rocker, he later used Venom rockers which have no offset but inboard adjusters. Much of the improvement in performance is, he admits, due to visits to the dynamometer at LEDAR where Leon Moss with his tremendous experience of modern racing engines was impressed with the performance of the 250 when it pulled 29 bhp at 8,500 at the back wheel. And he was thinking it was a 350!

Dave maintains that it would be possible to get the same results even-

Almost indistinguishable from its big brother, the Holmes MOV is nearly as fast and often won unlimited capacity Historic races run by the Vintage MCC. Close up shows wired preparation for racing

tually by the road testing he employed before but admits it would have taken a long time. A long time perhaps to find as he did on the brake, that rubber mounting the carburettor cut down the blow-back and enabled him to come down a lot on the size of the main jet.

A long time perhaps to dispel the widely held belief that with methanol you can't be too rich. Being too rich costs power without you being able to notice it... except on the brake.

To get the carburation spot on he even modified the needle of the TT10 carbs, tapering them off to a point.

There are, of course, limiting factors when you wring at least twice as much power out of an engine than the maker ever intended. Obviously there are mechanical problems, and I will deal with those later, but an unexpected one was the difficulty of getting a magneto to provide an adequate spark at such high revs and at such high compression ratios... well into the teens. To use an old tuner's expression, the internal pressure "blows the spark out". Dave now winds his own magneto armatures to get optimum voltage but still has to reduce the plug gap to way below the normal, and the situation is so critical that he can only use a Champion 54 R plugs, just one grade from the hardest, because they have sharp-edges electrodes which need marginally less voltage than makes with the more normal round-section electrodes. He did, incidentally, try a magneto wound specially to provide twin sparks and there was a slight power gain but it aggravated the spark plug problem.

Why, I asked, did he not do what most serious contenders have done and go coil or electronic while still apparently using a magneto for appearance purposes?

David Holmes with KTT Velocette on DPX2 dynamometer

"Because of the regulations. I did not want to give anyone a chance to complain", he replied.

Quite so. When you have a little ding bat of a pre-war two-fifty that beats up pre-war and post-war five hundreds so contemptuously, the eyes of the paddock are on you.

Mechanically, the limitations of some of the internals of a "cooking" 250 of the early 1930s are obvious, and rather than wait to see what will break, Dave as an engineer has made some new parts capable of taking the strain of twice the revs and more than twice the power of the original engine. Steel flywheels, a thicker pressed-in crankpin with a caged needle roller big end not unlike the later Velo bearing support a new connecting rod which is a masterpiece of machining. Two ribs strengthen the big end and little end eyes, in best racing practice. Pistons, again "home made", are another work of art being somewhat like a Manx Norton piston in miniature. Velocette's cost cutting stratagem of turning a 250 into a 350 by lengthening the stroke to a length that would now be looked on with disbelief obviously wouldn't do if the 350 was to be faster than its 250 brother. New flywheels and connecting rod have brought the dimensions down to the more ideal 75.8 bore and 77.1mm stroke, to give 348 cc.

Some more Holmes engineering has produced semi-close internals for the standard 250 box (using Venom gears slimmed down, and the 350 has modified Venom TT ratios). The 250 pulls 6.5 to 1 top, and the 350 5.5 to 1. The clutches use bonded linings on home made plates but are otherwise normal except that the 250 now enjoys a toothed belt and a home made clutch sprocket gear wheel in light alloy.

"I was near the limit with the primary chain which would only last a meeting. They said you couldn't fit a belt to a Velo... not enough room and when the clutch tipped as it was freed the belt would come off. Those sort of stories. Well, I've done it and it works. It is inside the same case I used for the chain and no one would have known if they had not seen me changing it in the paddock."

"But why not start with a proper racing engine instead of a 'cooking' roadster?" I asked.

"Because the late MAC crankcase is stronger." he replied. He must have sensed my disbelief so, producing an MAC drive-side half crankcase and a Mk. 8 KTT equivalent, he demonstrated with the aid of a craft double calliper arrangement that whereas the MAC crankcase is a full .23in thick the Mk. 8 is slightly less. Its ribs make up the thickness but the wall doesn't. The Venom crankcase, he says, is nothing like a strong as the MAC partly because of the greater overhang of the cylinder base flange.

So the crankcase which is often the limiting factor in developing a vintage type engine is perhaps the strongest type available, and I am sure the challenge of taking a roadster engine and beating racers with it appeals.

David Holmes works in a Nottingham engineering works but builds his racers in a workshop in his parents' garden. He gives much of the credit for an almost unbeatable run of success in historic racing to his young rider Stephen Tomes whose father often followed him home... at a distance on an inordinately fast 250 Royal Enfield. Stephen Tomes later rode modern machines and David Holmes, who now has his own dyno, prepares a variety of racing engines.

D.W. Holmes, The Leys, Normanton on the Wolds, NG12 5NU.

Perhaps not the photogenic side of a Velo, but nonetheless interesting. Primary chains could not stand the 8,000 plus rpm of the Holmes MOV. Experts said the peculiar Velo clutch ruled out the toothed belts widely used by Historic racers. Holmes proved it could be done. Air scoop cools belt and clutch. No more oil leaks, either. Note resited oil tank to make room for carburettor. The crankcase is an MAC, a post-war one. More readily available and stronger on the drive side than an MOV crankcase

"A magnificent engine. But as for the roadholding..."

Stanley Woods was the rider Velocettes needed to assess the TT machines

The 1935 Mk. V KTT

I have thrown out enough hints already for readers to realize that my researches into the Velocette fortunes in the Island in the early 30s led me to believe that Percy Goodman and Harold Willis were barking up the wrong tree in searching for more power when all the time they were losing out on handling. The trouble was that their team men were of the old school, brought up on vintage machines which had to be fought round the twisty bits, men whose pride would never let them blame the handling of a machine for any deficiency in their lap times. And not ever having ridden anything better, they had no yardstick by which to truly assess a Velocette. What they really needed was a star rider who had ridden more advanced machines, to provide a yardstick, and who had the analytical skill to interpret his findings. In the way that many years later Duke and Surtees were able to assist the designers of the machines they rode. It so happened that the very man was ready and willing to serve Velocettes in 1934 but they turned him down flat for reasons which appear to be more personal than politic. When I say the volunteer was no less than Stanley Woods you can share my amazement at unearthing this piece of hitherto untold history. But I had it direct from Stanley himself and as I trace the story of Velocettes in the Island in the later 30s I now insert some glimpses behind the scenes provided by the great man himself. It was after the 1934 TT when he had finished fourth in the Lightweight on a Guzzi and had retired his Senior Husqvarna that Stanley turned his attention to Velocettes.

"I realised that a Velocette was my only hope of success against Nortons in the 350 class. I approached them at the 1934 motorcycle show but my suggestion was turned down... much to my surprise."

His surprise was understandable for any other manufacturer apart from Norton — who under Joe Craig had built up a team organisation which had no place for a "loner" like Woods - would have gone down on bended knees or waved bags of gold to have Stanley. The trouble was that Stanley was the true professional only interested in a machine with winning potential. He had to be. He was not riding for fun or for the security of a retainer. His business was to win races and to do that he had to pick winning machines.

If he was surprised at being turned down flat in 1934 he

was even more surprised in 1935. The Velocette team men for the Junior were Wal Handley, as first string, supported by H.E. Newman, the brilliant amateur, and Ernie Nott, the old campaigner. Newman had demonstrated his form by winning the Junior Donington Grand Prix — a pre-TT try out — on a standard KTT bought "over the counter" from L. Stevens Ltd. who were leading Velocette agents, but soon after he landed in the Island he went down with pneumonia. Stanley promptly popped the question again.

"Towards the end of practice one of their riders was hospitalized. I approached Veloce Ltd. once more, and once more I was turned down. As I was strongly fancied for a win in both 250 and 500 cc races [he won both] on the Moto Guzzi their refusal of my services free of charge surprised me more than a little. Later that day, talking to Joe Craig, I was told that their attitude was dedicated by Wal Handley who had refused to ride for them if I joined the team. I found this difficult to believe so I approached Mr. Percy Goodman once again and he admitted that this was the case. I pleaded with him for some time and finally left him late on Friday evening telling him to be sensible and to have a machine for me to practise next morning so that I could qualify."

"Next morning I went to the practice with the 250 Guzzi and hung around for quite a while until it was obvious that there would be no Velocette. I went off for a final lap on the Guzzi and on completing this learned that Handley had been involved in some sort of accident and would be unable to ride. [Handley had caught his hand in the rear chain while adjusting his brake on the move.] Within minutes I was asked to take over his machine. With time for only one lap I agreed to try the machine and qualify. I completed the lap but decided not to ride it as I had such good bikes for the other races, and felt that the Velocette was just that little bit different from the Moto Guzzis and I would probably not be a threat to the Nortons until I had completed two or three laps at racing speeds. That was a chance I was not prepared to take."

Earlier I suggested that Stanley Woods was the kind of rider Velocettes needed to assess their machine and help to sort it out. This was his finding after one practice lap - not a particularly fast lap at 73 mph but only 5 seconds slower than Crasher White who topped the list throughout on his Norton. On the last practice session, the favourites preferred not to risk breaking their models before the weigh in. Both Handley and Nott had previously lapped at 76 mph

"In particular," Stanley Woods said, "I found the engine magnificent, an almost certain winner. The gearbox ratios

Close-up of the Mk. 5 KTT shows the low-pressure oiling system which had a direct feed to the cam box via the large external pipe to the rear of the camshaft drive, the bevel gear and drive being no longer under pressure. Through bolts were used to retain the barrel and head. As mentioned, this was the last stage in Percy Goodman's "vintage" engine, with horizontal inlet tract

were quite unsuitable for the TT course... too low and widely spaced. The brakes were, by Moto Guzzi and Norton standards, almost non-existent and the roadholding, to put it mildly, was most uncertain on high-speed bends."

There it was in a nutshell. The lesson that Veloce had not hitherto learned because no one before had a suitable yardstick and such analytical sense.

The ice had been broken, however. Fate had taken a hand by eliminating Handley from a difficult personality deadlock at a crucial moment, and the foundation was laid for an association between Stanley Woods and Velocettes which was to have far-reaching effects. It was to reshape both the Velocette racing machines and their Isle of Man fortunes, and after some early misfortunes to give Percy Goodman the prize he wanted most of all. Victory again in the Junior TT.

But back to the 1935 TT series which, from all points of view, was to be a disappointing one for the Velocette camp. The loss of Handley and Newman before the Junior began was a real blow but it brought promotion to "works" status to Ernie Thomas, who took over Handley's machine, and to a rising star, D.J. Pirie, who seized the opportunity with a fist-full of throttle to outspeed all the other Velocette riders and come home fourth at 77.69 mph, albeit over two minutes behind the Norton third placeman, Crasher White. Only a few seconds separated the Velocette team, Ernie Nott being fifth, 19 seconds behind Pirie, and Ernie Thomas

37 seconds behind him. Les Archer, tipped as a Velocette favourite, was ninth, Noel Christmas 11th, H.C. Lamacraft, a lifelong Velocette private owner enthusiast, 12th, Arthur Tyler 14th. Billy Tiffen, better known later as a trials man, was 16th, Norman Gledhill 17th and A.C. Kellas 18th.

It was noticeable that apart from the works Norton one-two-three of Guthrie, Rusk and White and that of J.G. Duncan in seventh place (probably a semi-works machine) there was not a single Norton among the finishers. It seemed that Nortons concentrated on building a few pukka racers but that the over-the-counter models were not competitive. Velocettes, on the other hand, produced a considerable number of KTT models which were good enough to lift a competent private runner to the leaderboard.

The 1935 Junior, therefore, followed the pattern that had come to be expected. Victory for the illustrious Norton works team and an impressive demonstration of high-speed reliability by Velocettes, and particularly by private owners on over-the-counter KTT models, mostly the Mark 5 model which looked, and probably was, exactly like the work's mounts apart from the engine. For 1935 Velocettes had adopted a policy of standardisation, or rationalization as it would now be termed. The racers used the frame (less the odd surplus lug) which had been developed through racing and housed both the new roadster Mk II KSS engine or the newly introduced 500 cc MSS engine. It was longer in the wheelbase than the old vintage-type diamond frame used with added lower torque stays on the Mk 4 — perhaps too long for racing purposes, for high-speed handling was certainly no better and perhaps even worse than that provided by the earlier models. The front forks still had the characteristic bracing struts but the vintage-type Webb friction dampers with flat adjusting wings had given way to the roadster-type single damper on the off side with a milled handwheel or flat thumb nut for adjustment. The whole machine looked very much like the production KSS model which was to appear at the next Show.

The Mark 5 engine was the final development of the original Percy Goodman design of 1925. There were only two significant changes from the Mk 4 engine. The first was the use of what has become to be known as the "low pressure" oiling system, something of a misnomer because there was still high pressure to the important bits like cam box and big end but this time the pressure feed to the cam box was by means of an external pipe and no longer was the entire camshaft drive of shaft and bevels under pressure. They now relied on oil draining down from the cam box. The other alteration was an extension of the crankcase mouth to provide better support for the barrel and the use of through bolts to hold head and barrel down.

The 1935 "works" models in both 350 and 500 cc form used the old "dog kennel" cam box engines reworked a little and given big-fin alloy barrels — an alloy muff cast on to an iron barrel. Megaphone exhausts were used for the first time.

If the Junior had been a little disappointing for Velocettes the Senior was disastrous for in the event only one 500 started. Ernie Thomas was on the machine Newman would have ridden, and he only got as far as Ramsey hairpin before he was eliminated by a broken fork spring. Handley, of course, was eliminated by his hand injury, Newman was in hospital and Ernie Nott did not turn up on parade. This was the year when fog caused cancellation of the Senior on Friday and threw the Island into a turmoil. Thousands of trippers slept out that night hoping that the race would be run on the Saturday. Thousands more had to go home bitterly disappointed. On the Saturday morning the Island was still

The Mk. V KTT as used by private entrants in the 1935-36 period. The frame was similar to the roadster Mk. II KSS and MSS models, giving a longer wheelbase than the earlier KTT models. The complete machine bore a close resemblance to the Mk. II KSS model

Wal Handley on the works Junior machine in 1935. Towards the end of the practice period he caught his hand in the rear chain when adjusting the rear brake on the move. At the last moment of practice the machine was offered to Stanley Woods who completed one practice lap on it but decided not to ride it in the race. Ernie Thomas rode it in the race, eventually finishing sixth

wreathed in mist and it seemed unlikely that racing would be possible. Some of the riders had races on the Continent on the Sunday and left on the morning boat, among them the Velocette team man Ernie Nott.

The next important race was the Dutch TT and Stanley Woods came to the line for the first time on a Velocette. In his words: "I agreed to ride it [probably the ex-Handley machine] in the Dutch TT and after breaking the lap record in practice I led the race until I had to retire with lubrication trouble. Following the Dutch TT. I agreed provisionally to ride for them in 1936 provided drastic changes were made in the design, and I later visited the works to finalize plans. I spent the whole day with them and went into every aspect of the proposals except finance.

"Finally as the day drew to a close I raised the point and they were quite surprised. I think they had some idea that a rider hibernated during the winter or possibly lived on air. However that difficulty was resolved and an interesting and reasonable successful four years followed. I should have won the 1936 Junior but the camshaft drive fractured. I could have won the 1936 Senior if a mystery misfire had not developed towards the end of the sixth lap. In 1937 Nortons sprang a surprise with a machine that had gained a fantastic amount of speed over the winter, but in 1938 and 1939 I managed to get home ahead of them in the Junior while in the Senior I had to be content with a hard-earned second place in 1936, 1937 and in 1938, while in 1939 the BMW and Nortons were too good for us."

Came the Ulster GP of 1935 and the usual triumph over Nortons in the 350 class, Wal Handley, now fit again, winning with Ernie Thomas second. For the 500 cc class a couple of 500 models were loaned, one to Arthur Tyler who had finished 14th on his privately owned 350 in the Junior TT, and the other to a local policeman, J.J. O'Neill. Tyler, in what was to be his last race before he retired, rode the race of his life to finish third behind Milhoux on the five-speed Belgian FN, both being quite a way behind Jimmy Guthrie on a very quick Norton which averaged over 90 mph From Arthur Tyler I was able to obtain some rider's impressions of a privately owned Mk 5 KTT and the works 500 cc machines.

Having signed up to ride Vincent HRD machine in the 1935 Senior, Tyler put a KTT on order for the Junior. He was not quite sure what prompted him to choose a Velocette, after scoring a number of successes at Donington on Norton, but thought he must have been influenced by the number of privately owners Velocettes which always won replicas in the Junior, suggesting that as an over-the-counter machine the Velocette was a good buy. With Nortons it was not so simple. The work's machine were very good but unless you had a lot of influence it was difficult to buy a quick one. He knew this from personal experience because he had once been able to buy a "Joe Craig" motor intended for someone Joe Craig knew and it had been very quick... leastways until a gudgeon pin broke within a few yards of the finishing line at Donington. After it had been back at the works for a rebuild it was never as quick again.

His KTT, one of the first Mk 5 models, was delivered in time for him to ride it in the Grand Prix meeting at Donington in May, a meeting with 100-mile races intended as a workout for the TT. In the 350 race Tyler took the lead in the later stages when Harold Daniell had fallen off his works AJS but was eventually beaten by a few second by H.E. Newman. Tyler was lying third on the last lap in the 500 cc race on his 350 when he ran out of petrol due to a leaking tank. Not a bad work-out for a same-as-you-can-buy 350.

His ride in the Junior TT was made uncomfortable and rather dodgy because of a trifling fault. The rubber bush locating the new-type fork friction damper (same as the roadster models) broke up, which meant that the damper was useless and the forks hopped up and down unchecked. After riding Nortons he was never completely at home on the Velocette. The steering, he thought, was a little too light, the gear ratios were not close enough, and the change (down for down) awkward. And he didn't like *that clutch* which made push starting tricky. But the engine was very good.

Apart from the tuppenny ha'penny fork damper bush which disintegrated, there was only one fault on the machine at the finish. The primary chain was worn on the sideplates because the inserts had come loose in the clutch plates and allowed the sprocket to wander. Chain lubrication on this model, incidentally, was by means of a pipe taken from the oil tank at a fairly high level for safety's sake, with a 300 Amal jet in it to meter the flow.

The 500 he rode in the Ulster was a very different proposition. It was the first time he had ridden a machine with a megaphone and megaphonitis, and he recalls it was very difficult to ride it to the start through Belfast. The power came in with a rush and it was advisable to have the model pretty well upright when it did come in. Although he was of course very flattered to have been loaned this very exclusive piece of works machinery he did later begin to wonder if his good fortune was due to other people not being

Billy Wing: Velo agent

FROM BEGINNING to end, as I have said before, Velocettes were made by a family, the Goodman family, and over the years, particularly before the last war, many of the dealers who sold Velocette became friends of the Goodmans and part of an even greater family. Most early members of this close knit manufacturer-agent family are long gone but one of the best examples of the unique relationship which grew up, in his 90th year, recalled how Velocettes changed his life.

W.A. "Billy" Wing sold cycles and mended punctures in a tiny corner shop in Daybrook, a suburb of Nottingham, in the early twenties and gradually got in to repairing motorcycles and riding a Big Port AJS on local grass tracks. A fearsome ex-Brooklands 500 Ariel followed (of which more anon) and as his reputation as a local racer grew so did the motorcycle side of his business. The day his business changed was around 1931 when George Denley (then Velocettes, only salesman) dropped in out of the blue and asked him if he would like the Velocette agency for Nottingham and County. The previous agent had apparently gone broke.

"I said I couldn't afford to stock many machines but he said they couldn't make many machines anyway and I could take one machine a year that would be alright". So Billy became a Velocette agent, a rider agent like most agents of the period, and his enthusiasm for the marque soon resulted in sales. One of his early customers was local girl Molly Twigg who soon became a jolly good trials rider on a cammy model (after she married Triumph specialist Alf Briggs she naturally switched to Triumphs and rode Trophy models in the ISDT when not grass track racing). Billy rode his Ariel special (I say special because it had been fitted with Castle bottom link forks by someone trying to curb its wayward ways) at the first Donington meetings but soon swapped over to a Mk 1 KTT Velo, as befitted a Velo agent. "I really wasted my first years at Donington because the Mk. I was not

Billy Wing in the IoM in 1935. He had won a free entry in the MGP

too keen to take it on. For it was apparently a bit of a handful and his race was not made any easier by the long mattress dualseat... the "Loch Ness Monster" seat. The trouble was that there was nothing to stop him gradually slipping off the back, which meant he had to constantly haul himself back on again by pulling on the bars, which did not help the straight-line navigation. The 500 was fast — well over 110 mph — but then again Guthrie's Norton was much faster and easily passed him on acceleration through the gears. He saw the engine dismantled and does not recall anything unusual about it except that the valve-included angle seemed less than usual and the head rather flat. The Ulster was Tyler's swan song for he retired immediately afterwards to marry and devote himself to the family business.

If 1935 had not been a very successful year race-wise it was one of the most significant in Velocette history for it was the year of truth as far as handling characteristics were concerned and the opinion of no less a counsel than Stanley

---

really powerful enough but my fortunes changed when I got a Mk 4. After that I began to do quite well."

I know he did because in those days I was a reporter covering the Donington meetings for local papers and was constantly writing the name Wing in "local rider wins the big race" pieces. And I came to know Billy W. and his little corner shop because I tried to buy the Ariel special from him and now count myself fortunate that I only had £12 instead of the £18 he wanted for it, for assuredly it would have shortened my life terminally if I had ridden it, I like to think that realising the Ariel was, to say the least, several sizes too big for an undersized teenager he did not try as hard to make a deal as he might otherwise have done. It's a small world, but a week before I wrote this memory the aforementioned Alf Briggs showed me a photo of the old Ariel, still with its Castle forks but in trials trim with a sidecar, in action in the fifties.

Although Billy Wing was a regular placeman and occasional winner at Donington he was not so fortunate elsewhere.

He in fact won a free entry at Donington for the 1935 Manx GP through a placing in a Donington meeting but was caught out by a patch of mist on the Mountain on the first day of practice, hit a bank and damaged himself too much to make the race.

Jumping on to 1938, he was allocated one of a batch of "special" Mk 8s produced for selected riders like Archer, Newman, Pope etc. He couldn't remember what was special about them but thought it was to do with the valve gear (Peter Goodman says it was a special cam and an extra scavenge pump below the bottom bevel gear). With this 'hush hush' model Billy had high hopes in the Ulster but when he came to the warm up his engine it smoked profusely.

"Willis said pull in after a couple of laps and we will have a look at. I did and they said it seemed OK but it cost me a couple of minutes all told and dropped me to 9th."

In fact Billy had his share of bad luck racing. Like when he led a 40 lap race at Donington and really thought the £75 prize money (a great deal of money then) was within his grasp. Then, with three laps to go Harold Daniell, trying too hard fell off and his spinning bike knocked Billy off.

"I picked my bike up and it was a mess. One handlebar was bent right down beside the tank and the clutch lever was hanging down. Somehow I got it going and finished fifth behind Newman, Tyler, Watson and Booth. When Craner saw the bike at the finish he said 'You've never been riding it like that' When I said I had he said 'Well, you wouldn't have done if I had seen it".

Back to business, Billy sold all Velocette models but could never get enough. Things improved when the M range became available but the LE nearly broke his heart. "They should never have made it, it gave a lot of trouble. I sold them to the local police and had Tommy Mutton (works mechanic) staying with me for a week while he changed all the clutches. As for the scooter...."

Billy had long ago become a family friend of the Goodmans, visiting their homes and having lunch at the factory (lunch at the factory in the Boardroom which doubled as the executive canteen was always a social affair. I was a bit intimidated by the faded grandeur of it, the panelled walls and all when once I was entertained to the modest fare which was the norm).

He remembers well the day in 1939 when Eugene showed him the Model O and got him to try it out. "Wonderful bike. I got it up to 90. Now if only they had made that instead of the LE"

Eventually, in the post-war years Billy, who never gave up his corner shop, gave up his franchise for the County but retained his Nottingham agency until he retired in 1966.

Woods served to galvanise them into action. A party including Wal Handley and Walter Rusk, as test riders, went over to the Island in the autumn with Harold Willis and Charles Udall as technical experts to try and lay the steering bogey once and for all.

Of this fact-finding mission Charles Udall recalls: "Up to 1934, when 80 mph was quite fast, the steering was perfectly satisfactory and the lessons of racing were built into the standard machines. After 1934 the faster men, like Handley and Rusk, were beginning to get into trouble on high-speed bends. When we went over to the Island to try out the machines Walter and Harold were busy cutting out engine plates to move the engines farther back. Walter's idea was they wanted more weight on the back wheel to keep it down. After I had seen Rusk in mid air at the 13th Milestone I came to a conclusion. I said to Harold that I didn't know whether he agreed or not but I thought he was on the wrong lines in moving the engine back. Harold said he was backing Walter Handley so I went to see P.G. [Percy Goodman] and told him my idea. He said it seemed reasonable. You see, I had always been interested in aircraft design and one of the first principles of directional stability in aircraft is that the centre of air pressure must be behind the centre of gravity. It seemed to me that a motorcycle with its wheels off the ground was very much like an airplane without a tail fin to stop it turning right round and we needed the centre of gravity well forward of the centre of pressure to make it stable."

Later Stanley had two test sessions in the Island on the new experimental model which resulted, and of this he writes: "I tried 'my suggestion' frame, and Percy Goodman's 'agreed idea', on two occasions in the Isle of Man during the winter of 1935-6 and we settled after considerable discussion on my suggestion re geometry, etc."

The result of all the work and experiment during the winter was revealed to a rather startled world of race enthusiasts in the Isle of Man the following May (1936).

It may not be startling to modern eyes, for pivoted-fork rear suspension with a simple tubular fork - tapered tube then, as on all Velocettes - and almost vertical suspension units, supported by a triangular rear-frame section, became the norm. In 1936 there had never been anything quite like it before. Other pivoted-fork designs, the Guzzi with springs under the engine and Vincent HRD with springs under the saddle, had triangulated multi-tube forks. Nor had anything more sophisticated than friction dampers been thought of, usually tightened up pretty solidly so that the frame only reacted to the worst bumps and for the most part behaved like a rigid. But here, with the oil-damped units supported by air pressure, was great sensitivity permitting the rear wheel to follow the slightest irregularity in the road yet checked constantly by necessity of forcing oil through a jet. These legs were adaptations of oleo struts used on aircraft undercarriage and already developed to a high pitch. There was one great advantage over spring units. The "strength" could be easily adjusted by pumping up the static pressure... simply a matter of blowing them up with a tyre inflator. And the internal function was attractive because as the legs were compressed and the piston displaced oil through a jet it forced it into the air chamber, thus increasing air pressure and resistance to movement.

The purist in Percy Goodman insisted that this suspension should be called a shock-absorbing frame not a spring frame because of course there were no springs.

I have no recourse to describe the 1936 frame which was to house 350 and 500 engines for Stanley Woods in great detail, or to search for long-lost photographs, for it was identical in all but the smallest details to the well-known Mk 8 KTT of 1938-49.

In fact the only changes made before it went into production was the replacement of ball bearings in the fork pivot by bushes and the change from cable to rod operation of the rear brake.

As for the 500 cc version, there was no surprise for the engine was a three-year-old "dog-kennel" unit, but there was a bit of real hush-hush in the 350 motor for it was, believe it or not, a double-overhead-camshaft job which Willis promptly called a "double knocker" — and that is how the universal euphonious expression came about. The dohc engine, a one-off for Stanley to ride in the Junior, was a logical development of the "dog-kennel" unit with its short pushers for direct attack on the valves. Instead of the enclosed rockers there were two camshafts driven by intermediate idlers from a central spur gear. Practice did not reveal any spectacular improvement in Velocette times attributable to the spring frames — sorry, shock-absorbing frames. If the riders had an edge, they were keeping it dark.

The Junior was sensational for two reasons. First because Stanley Woods, very much the dark horse tipped for a runaway win, was out of the race before Sulby when the vertical camshaft drive sheared. Looking at these couplings used on all ohc Velocettes, I have always marvelled that they stand up at all. They seemed altogether too flimsy and bedevilled with sharp changes in section to withstand the constant reversals of valve mechanism but I must admit that I have never seen a broken one and neither had Velocette. It is not, therefore, surprising that the failure was attributed to the extra load of the dohc gear. It seems remarkable that such a promising design should be put on the shelf after one isolated failure of a small part and I am inclined to think that the decision to revert to a single camshaft was dictated by policy — it being too expensive for quantity production — or through dudgeon at its infuriating failure. Perhaps a little of both. With Stanley out, the reputation of the marque lay in the hands of Ernie Thomas and Ted Mellors... Mellors, incidentally on a rigid-frame machine, virtually a prototype of the subsequent Mk 7 KTT, which can be most easily described as a rigid version of a Mk 8. Apparently Mellors still preferred a rigid frame. But Ernie Thomas was riding brilliantly in what must have been his finest hour for a long time. For four laps he clung gamely to the leading Nortons of Guthrie and Frith, conceding but 20 seconds to Frith and 53 seconds to Guthrie. Mellors was backing him up but two minutes behind. Sensation number two came on lap four when Guthrie had his rear chain come off at Hillberry. Told he had outside assistance from a marshal in push starting, the stewards exclude him from the results when eventually he comes home. Later they change their mind and give him the money for the second place, it is presumed he would have won if he had not been stopped by officials. An extraordinary decision which was difficult to understand. While Guthrie was refitting his chain and struggling to get going, Thomas for a brief moment jumped into second place only to come unstuck at Quarter Bridge and lose some time kicking things straight and lose even more time when the stewards had him stopped next time round for a safety check which dropped him to fifth place, while Mellors moved up behind Guthrie.

Frith won, J.H. White, surprisingly restrained, was second, Mellors third and Thomas fourth. What a sensational race it was.

Further research into the Isle of Man photograph archives of S.R. Keig suggest that far from there being only one "double-knocker" Velo in the 1936 Junior, the one which let Stanley Woods down when the camshaft drive failed, there were two others ridden by Mellors, who finished third, and Thomas, who was fourth. One would have thought the maker would have been pleased to state this in vindication of the design but perhaps the decision had been made to drop it and the least said about it the better.

By comparison, the Senior was uneventful until the last lap. Guthrie led all the way while Stanley Woods on the airsprung Velo played his characteristic lone wolf role shadowing him until the last lap when he intended to pull a shattering last lap out of the bag. A war of nerves between Stanley, with his private signalling station and his own secret tactical plan, and Joe Craig controlling his team with the precision of a guards RSM The drama goes like this... Stanley stalks Guthrie for three laps, not too close for fear of bringing out the "go-faster" signs, and then gradually closes the gap to 22 seconds, helped by the fact that Guthrie had to make a precautionary second pit stop for a wee drop of fuel and was baulked in doing so by team-mate White. Woods with enough fuel to finish shows his hand with a record lap at 86.89 mph and Guthrie is told to go flat out. He started ahead of Stanley who now has his chance to spring his final surprise. He fails by a mere 18 seconds. On the fateful last lap when Stanley rang down for full steam ahead the big Velo with its three-year-old engine in its super modern frame developed a misfire... plug, perhaps. Stanley had the consolation of the fastest lap and everyone who knew about the misfire was pretty sure he would have won but for that. Quite a way behind came Freddie Frith and Crasher White and Noel Pope, better remembered at Brooklands but no mean performer on the road. Believe it or not, 350 Velos were sixth and seventh, Chas Goldberg and Billy Tiffen being the cheeky upstarts. Ernie Thomas having got

Arthur Tyler on the works 500 he rode into third place in the 1935 Ulster GP, his last ride before retirement. Very tall... a gentle giant of a man... Tyler dwarfs the Velo as, conscious of his responsibility, he concentrates on a slow bend

up to fifth place on a 500, went out on lap four with carburettor needle trouble.

Mellors went on to win the Belgian GP and was second in the 350 class in the Swedish GP and the Ulster.

In some of these latter races Mellors did, I am told, use "the Little Rough 'Un" — contrived by Willis from odds and ends — a roadster Mk 2 KSS alloy head on a Mk 5 bottom end, which was used by Austin Munks to win the Manx, but as it was never released to the press it is not mentioned in reports.

The spring-frame models had come through with flying colours. Everywhere the knowledgeable had been impressed with the way the bumps were smoothed out. Velocettes were in the ascendancy again and all that was necessary now was more power to match the new-found road-holding.

Mark 6 KTT mystery solved ?

EVER SINCE the story of Willis going into the stores and collecting a Mk. II KSS head and other bits and pieces to build a new top on a Mk. KTT bottom got out, it started a hunt which has engaged Velocette researchers all over the world. At first it was a simple "where is it now" hunt for the bike nicknamed by Willis as 'The Little Rough 'Un' which gained a certain immortality as it took Lincolnshire's Austin Munks to victory in the 1936 Manx GP. A hunt heightened because we had been told that it was the only Mk. 6 KTT ever built and a chase which had me following up clues in many places. When Dave Davies, a Vintage Club friend, also hailing from Lincolnshire produced photos of his friend Clifford Ellerby racing a bike which at a distance looked the split image of the one Munks had ridden, I felt sure it must have been the one and only Mk 6 'Little Rough 'Un'. Especially when Davies recalled that Ellerby had bought it in the early fifties from a man in Manchester name of Binder. I remembered the ad; in fact I found the cutting which said "Genuine works Mk. 6 KTT Velocette". I think the price was £45, don't laugh, that was a lot of money then. Binder was obviously Franz Binder, the Austrian who came to England before the war and worked at Veloce as a tester and

Austin Munks rounds Governors Bridge on 'The Little Rough 'Un' during his winning ride in the 1936 Junior Manx GP

rode a Mk. 8 in the Island fitted with the friction suspension units as used on the "O". It was still quick too, for Ellerby held the lap record for the "old" circuit at Cadwell on it. However the story did not stop there. Ellerby had apparently sold it to an R.A.F. type who had ridden it at the Brighton Speed Trials. It had blown up with a broken con rod, it was said, and had afterwards reappeared on the grass. Bike restorer Ken Bake was certain he had ridden it on the grass but it had vanished again. Latest report is that it is in good hands in the South of England and is being restored.

This seemed to be the end of the story, but then researches by Velo experts Dennis Quinlan in Australia, Dave Masters and Bruce McNair in England came up with the revelation, backed by work's records and photographs, to prove that far from their being only one 'Little Rough 'Un' Mk 5 - KSS Mk. 2 hybrid, there were actually three more Mk. 5 - Mk. 2 KSS hybrids; what's more they were ridden in the 1936 Junior TT by Billy Tiffen Jnr., who retired with a broken fork spring, by Roger Loyer, the French rider who retired with lubrication trouble, and Harold Newman who finished 10th. This information allowed closer study of photographs of Munks on the 'Little Rough 'Un' at Signpost Corner and Ellerby taken at Cadwell from a similar angle. It became clear that they were not the same machine. Munks' bike had plain Webb forks of Mk. 2 type but with friction dampers each side whereas the Ellerby bike had braced Webbs. There were slight differences in oil drain pipes as well. It seemed that a later story that Frank Mussett had taken the Willis-built back to Australia with other race bikes might be correct. The question that troubled me was why a firm so scrupulously honest in other ways should have dissembled in this matter by saying only one Mk. 6 KTT had been built when they in fact built three others as like as peas.

The answer, I suggest, is disappointingly simple. *The firm did not say anything officially* because there never were any Mk. 6 KTT models. Before you raise your eyes in disbelief let me make a distinction between Works racing machines, which could have any specification they chose at the time and were no concern of anyone outside, and catalogued KTT models which were built to a standard specification and offered for sale to the public. I have no doubt that it was intended that there would be a Mk. 6 KTT, essentially a Mk. 5 with a new square fin alloy head and probably a frame modified to bring the engine farther forward. When the head castings were late arriving, Willis, as we know, 'cobbled up' the 'Little Rough 'Un'... he was too polite to call it a Bastard as another man might have done. It worked well enough for the firm to build three more for works supported riders Loyer, Newman and Tiffen. Not quite what had been intended but somewhat better machines than the now obsolete Mk. 5.

After the TT it was obviously decided to miss out Mk. 6 from the sequence of KTT models for sale to private owners and when the big heads eventually arrived and had been developed to their satisfaction they called the result the Mk. 7 KTT. The front down tube had certainly become more upright in accordance with the findings about centre of gravity resulting from the need to satisfy Stanley Woods' standard of steering.

So where did the 'only one' story come from? From Bob Burgess, I am sure. It was a lovely behind-the-scenes story from his recollections of days at the factory, typical of the informality which allowed Willis to go and help himself in the stores (issuing a requisition countersigned by some high panjandrum and then dispatching a minion to collect the parts). As I say, a lovely story which I seized on with relish. The only flaw was the use of the letters KTT.

The 'Little Rough 'Un' and the three lookalikes were not KTT models at all but factory specials. Bob Burgess was Service Manager and would not have much contact with the racing department. He may not have known of the three other Mk. 5 - KSS Mk. 2 specials. Had he told us that only four Mk. 6 KTTs had been made it would not have been such a good story and set us all agog.

A good historian checks and double checks all information he is given. I am not that sort of historian, but I should have been suspicious when later first Ernie Thomas and then Peter Goodman expressed surprise when I told them that only one Mk. 6 had been built. Thomas said he thought he had once raced the Willis 'special' and the 'hot' bronze head MSS built for young Peter had, he was sure, a Mk. 6 frame. I am left with an awful doubt. Were there four Willis type Mk. 6 models as I have suggested, or only three? The works register shows only three:

June 1936 Veloce for H.E. Newman for TT KTT 621   Frame 6TT3
June 1936 Veloce for W.T. Tiffen Jnr. for TT KTT 622   Frame 6TT4
June 1936 Veloce for R. Loyer for TT KTT 623   Frame 6TT5

Clifford Ellerby at Cadwell Park

Was 'The Little Rough 'Un' one of these, and if so which one? My view is that like many experimental machines it was retained at the works and would not therefore need to be entered in the records. The others were leaving the factory and needed to be accounted for. Loyer, the French rider, bought his apparently and the others may have been sold. The fact that 'The Little Rough 'Un' was loaned for the Manx suggests it remained at the factory until given to Frank Mussett by Percy Goodman, with war approaching.

Austin Munks turned up a shot taken as he rounded Governor's Bridge on his winning ride in the 1936 Manx and from this it is just possible to establish that the machine was, apart from its KSS head engine, a more or less normal Mark 5 model. He recalls that the tank was unusually deep and narrow — this is noticeable from the photograph — and was, he thinks, a forerunner of the Mark 7 and 8 tank. Other points which have come back to him are that the forks had a guided spring — again like the Mark 7 and 8 — and that the rear hub was cone shape. This follows, for conical light-alloy rear hubs had by that time been employed on the work's machines.

Austin Munks remembered "The Little Rough 'Un" well. "It was a proper little screamer, very easy to over-rev. Of course we hadn't any rev-counters in those days and it was very easy to go over the top. I touched the valves in practice. I remember it was very light and pulled a higher gear than normal .. I had a job to get a 25-tooth sprocket. I think the compression ratio was higher, too, and the cams were very comical." "Why was it called the Little Rough 'Un?" "I don't really know, but it certainly did vibrate. It gave me pins and needles and it had six pairs of handlebars at once."

He seems to think it had an unusual carburettor, but he can't remember what it was, although he does remember that it very tricky and temperamental to start.

Another shot of Austin Munks on the Little Rough 'Un in the 1936 Junior Manx: at Signpost

When the black and gold beat the black and silver

Nortons did not begrudge the Goodmans their overdue TT victory

The Mk. VIII KTT: masterpiece in metal

Having finally sorted out the handling problems by shifting the engine mass forward, with the added bonus on the work's models of a pivoted-fork rear suspension which was way ahead of all competitors, it was to be expected that for 1937 Velocettes would turn their attention to the engine department. The "dog kennel" engines, so called, as explained earlier, because of the gable end of the rocker box which fully enclosed the rockers while leaving the hairpin springs exposed to the cooling draught, were now decidedly long in the tooth, being three years old. The dohc version was in disgrace, having let Stanley Woods down so dismally in the 1936 Junior. It would have been but the work of a few minutes to have beefed up the Oldham coupling in the drive shaft which brought it to a halt on the first lap but Percy Goodman, for reasons best known to himself, decided to forget it. His decision might have been influenced by the fact that the dohc design was unlikely to become an economic proposition for roadster use and Velocettes *did* like to preserve a relationship between their racers and their roadsters. Although they raced for their own pleasure, for the prestige it gave to the marque and for the beneficial effect on sales, the Goodmans were firm believers that the underlying object of racing was to improve the breed of roadster machines. They were, therefore, dead against the manufacture of exotic racing machines which would never reach the public.

In the very beginning they raced tuned-up editions of their production sports models. Then as the gap between racer and roadster widened they produced the KTT line which, if not entirely suitable for road use (for the day when efficient mudguards and lighting equipment were considered essential), could be sold to the public and raced straight from the showroom. Until 1933 the family resemblance between the work's racers and the KSS roadsters was

Moment of triumph. The combination of the riding skill of Stanley Woods and the design genius of Percy Goodman in producing a spring frame many years ahead of its time finally resulted in Velocettes regaining mastery of the Junior TT. Here Stanley Woods and wife Mildred are seen in 1939 after winning the Junior for the second year running. On his left are Charles Udall (in leather coat), Percy Goodman (in lounge suit) and Eric Brown (white overalls), Stanley's long-time mechanic, and Tommy Mutton

marked, and although the "dog kennel" engine differed in the upper works they were in fact experimental models which might eventually have had some roadster counterparts and did share some features of the push-rod models, chiefly in the matter of port angles. But before these engines

Mark 7 KTT Velocette equipped for road use with a "bung" in the megaphone. Note steep front down tube. This is probably a prototype as it has square 'works' head and special front brake

reached the point of production as KTT models there had been significant advantages to casting technique which had made possible the one-piece alloy head of the Mk. 2 KSS model, with the whole of its valve-operating mechanism enclosed in the head casting, and this development made possible a greatly improved racing machine. Again the production roadster was contributing the basis of the racing engine, as it had done in the early vintage period. The work's racers had huge alloy heads of the same basic design as the Mk. 2 KSS models already in production but with rectangular troughs to accommodate the hairpin springs and greatly increased finning — square in plan view with quite sharp corners. The fins of the alloy cylinder "muff" were nearly as generous, almost hiding the vertical camshaft drive though shorn off flat on the drive side. Though later familiar today to enthusiasts accustomed to the Mk. 8 KTT models, the new engine with its massive head and barrel dwarfing the typically slim crankcase created quite a

In the Velocette camp at the Nursery Hotel, Onchan, Stanley Woods watches the preparation of his 1938 Senior TT machine. The oil pipe snaking up from the lower bevel housing is a take off for an oil pressure gauge used when setting up the oiling system

sensation when the first illustrations appeared in May 1937. Striking, too, was the "plumbing". The lubrication was much the same as on the Mk. 5 KTT with external feeds to cam box and bevels, but now oil return was entirely by gravity and consequently the drain pipes were generous in dimensions. Large bore metal pipe fixtures were taken from the valve compartments in the head, brought together in a Y junction discharging into a flexible hose leading to the crankcase. Few alterations were made to the cycle parts for 1937 although the one-piece saddle and rear pad had gone for good and the saddle appears to have been rigidly mounted on the rear sub-frame, no separate spring being thought necessary in view of the oleo suspension. A most cunning piece of detail design allowed the use of rod operation of the rear brake without the normally attendant see-sawing of the pedal under the action of the pivoted fork. Anyone unfamiliar with the Mk. 8 design can best visualise it having, in effect, two brake pedals, one — minus a foot pad — being pivoted on the fork and the other, the pedal proper, pivoted on the frame, just behind the fork pivot. The pedal proper was arranged to press down on the secondary "pedal" at a point in line with the fork pivot so that no relative movement occurred, yet conveniently arranged to the rider's foot.

The traditional strutted front forks now gave way to a simple girder design, superficially like the production heavyweight fork, but incorporating detail refinements, the most noticeable being a telescopic guide inside the fork spring pivoted on Silentbloc bearings at either end to ensure that the fork spring operated under ideal conditions, being free from the bending stresses inherent in normal designs. More than one machine had been put out of the TT by a broken fork spring and Velocettes, with their meticulous attention to detail, were to lay this bogey once and for all. So successful was this fork that it was used without alteration to the end of the firm's racing career in 1953. Work's models for 1937 had a conical rear hub in elektron alloy and a new alloy back plate for the front brake with a cast-in air scoop. The hub itself appears to have been of normal design.

The new Velocettes were seen in action for the first time in the traditional proving race, the North West 200, and showed their form in both races, Stanley Woods leading Guthrie's Norton for eight laps before he fell off, and Mellors leading 'Crasher' White in the Junior category until something went wrong. It was obvious that Nortons had not been resting on their laurels during the year, and in fact they had taken over where Velocettes had left off in producing dohc engines... remarkably like the ill-fated 1936 Velocette... TT practice showed that the Nortons had a slight edge — unless Velocettes were foxing — Guthrie putting in the best Senior lap at 87.82 mph, a record. Stanley Woods' best was 84.9 and Mellors (who still preferred a rigid machine) did 83.1 before casting himself off. Frith similarly set up a Junior record, of 82.6 mph which was a shade too quick for either Mellors or Woods, the former topping the Velocette list with 81.2 mph Stanley set up a record of sorts. On one lap he broke his gear lever at Ramsey and completed the lap in bottom gear to average 73.7 mph As it was in the practice so it was in the Junior race. The dohc Norton just ran away with it to a triumphant Guthrie,

Stuart Waycott and the ISDT 600

WAYCOTT HAD come up the hard way via one-day sporting trials. In 1935 he was a private owner on a 495 MSS sidecar outfit and was the only British sidecar man to finish the road section. He did not complete the final speed test because his sidecar chassis had broken. Henry Laird in his supercharged Morgan was the only British passenger machine driver to finish the event that year. But Waycott's performance in getting further than anyone else won him the coveted sidecar spot in the Trophy team for 1936. Velocette backed him to the hilt by loaning him a detuned 500 ohc TT engine.

I am told that at first a Mk. 2 KSS 350 enlarged to 500 cc was tried but proved no more potent than a good MSS.

That year we had an easy win through the Germans striking mechanical trouble, but it was obvious that we had no sidecar machines capable of matching the 600 cc supercharged BMWs. So for the 1937 trial in Wales, Velocettes enlarged the TT motor to 595 cc The trial and its needle match between Great Britain and Germany was finally decided at the speed test at Donington and here Waycott rode a well-judged "race", refusing to be flustered by the leading BMW but not giving much ground. Britain won the Trophy by 10 seconds thanks to real TT riding by Vic Brittain on an overhead-camshaft three-fifty International Norton.

For 1938 the outfit was rebuilt with the swing-axle spring sidecar chassis, and the oversize motor was tuned to give 37.5 bhp with a silencer. Ordinary sand cast pistons would not stand up to it and a forged piston had

Percy Goodman, chief architect of Velocette leadership through design, was the most critical tester of his own machines. From a family album is this snap of him on the ohc 600 cc outfit which Stuart Waycott rode to victory in pre-war ISDTs. With him on a prototype LE model is his daughter-in-law, Mrs. Maureen Goodman. Note twin headlamps mounted on legshields

Frith, White one-two-three. Guthrie put the lap record up to 85.18 mph and Frith equalled that, so if anyone was foxing in practice it appeared to be Nortons. Stanley Woods held grimly to fourth place all through the race, even ousted White from third place on the fourth and fifth laps, but Joe Craig, controlling his team with military precision and having at least a mile an hour to play with, re-formed his team for the classic finish. Mellors ably backed up Woods for the first two laps on his rigid model and then retired, to be promptly replaced in fifth place by Ernie Thomas, shot up from a first-lap ninth. Thomas was passed by Daniell on the sixth lap but hung on to finish sixth.

The Senior race began according to form with Stanley Woods on the big Velocette chasing Jimmy Guthrie, but ended with a sensational win for Freddie Frith and a gallant but disappointing second place for Woods. Guthrie, as was expected, set the pace at the start with Frith second and Woods third. On lap two Woods passed Frith and then went after Guthrie. Guthrie filled on lap three while Woods delayed his stop to lap four, and this brought down the gap between the first two men to 19 seconds for a time until Woods had completed his fill-up. Enough to rattle all but the coolest team manager. Perhaps it did, in fact, cause Guthrie to push his motor a bit too much for on the fifth lap the Norton ran out of compression and would not climb the Mountain. This was the moment Stanley Woods and the

---

to be made at the last minute.

A number of behind-the-scenes references to his work at Hall Green appear in Irving's autobiography published after his death. A massive book because he crammed more into his life than half a dozen ordinary men (including three marriages!) and you soon come to understand how his early engineering training in Australia formed the foundation for his ingenious and innovative yet essentially practical designs in the two and four wheel world. A wonderful life story if you enjoy a good read but of special interest to Velocette and Vincent enthusiasts and followers of car racing and rallying too.

A typical Velocette "now it can be told" story concerns the big bore 600 cc ohc I.S.D.T. outfit when it was tested at Donington Park. After putting up satisfactory times (Irving himself got it round at near racing sidecar record speed) the outfit failed to come round.

It was assumed that it had run out of petrol, an assumption not discounted by Irving, and the ACU officials who had come to approve its use in the British team went home. Actually it had stopped with a nasty noise because the piston had broken up.

There was no time to get a replacement cast but Irving remembered a forging left over from the Aspin engine and worked through the night to machine it to his design for a stronger piston.

The Waycott outfit was a favourite machine of the Goodman family and was in use by them until well after the war, Peter Goodman using it when he was on leave from the RAF. Phil Irving borrowed it to win a first class award in the ACU National Rally, packing 720 miles into the 24 hours. Fellow Australians Frank Mussett and Len Perry accompanied him on solo Velocettes and it rained heavily the whole time!

Turned out to grass! The ex-Stuart Waycott 600 ohc ISDT sidecar outfit at a Southern Centre grass-track championship meeting at Warminster around 1953-55. The massive square-finned head of the enlarged TT machine can be clearly seen. The engine was later used by the Kendall brothers for scrambling and then displayed for many years at L & D Motors, Velocette agents for the Bristol area (Photo: C.D. Thomas)

Velo fans had been waiting for. Into the lead went Woods, and Hall Green hopes ran high. But Joe Craig had another ace up his sleeve, the slightly-built youngster, Freddie Frith, shy and retiring off the saddle, a veritable "tiger" on it and one of the first, if not the first, exponents of the flat on the tank all the way riding technique (when he was not poised above to wrestle with the brute). Joe Craig gave him the "flat-out" signal. His response was instant and sensational. By the end of the sixth lap he was tying with Stanley for first place at 87.88 mph Running nearly 10 minutes ahead on the road, Stanley finished at 88.09 and then had the agonizing wait while Frith rode the finest lap of his life. On the limit, scraping footrests, brushing banks, Frith hurtled round to record the first 90 mph lap and win by 15 seconds.

Almost unnoticed, amid the uproar of the sensational finish, Ted Mellors came home fourth on the other works 500 Velocette and in eight place was Les Archer on a 350, with the result that Velocettes won the manufacturer's team prize.

Abroad, Mellors and Thomas continued to harry the Norton teamsters. Mellors won the Ulster and the Swedish GP, was third in the Dutch, third in the GP d'Europe and second in the Belgian. Thomas was third in the Belgian GP and third in the Swedish.

A by-product of the racing development of the square-fin, all-alloy racing 500 Velocette was to defend British prestige in another sphere - the International Six Days Trial. This was the special sidecar outfit built for W.S. Waycott. Velocettes had not been interested in trials since the early 20s when Eugene Goodman and George Denley and Fred Povey were regular competitors. In those days trials were an excellent testing ground for new roadsters. The course were long and gruelling, and being largely over public roads did undoubtedly improve the breed. As trials gradually turned into short mud plugs they developed a special breed of short-wheelbase, high-clearance competition machine but no longer did much for the ordinary roadster machine. Most manufacturers welcomed and encouraged the specialization. It was much easier to gain kudos and advertising fodder by building a few special machines and running a factory team of expert riders to collar the awards then to rely on the efforts of private owners or works-supported riders on standard machines. Velocettes, with the honesty of purpose which runs all the way through their history, did not support this get-advertising-quick policy. Loyalty to friends in the retail trade did sometimes persuade them to help a little but never to the extent of a change of policy.

So Velocettes never marketed a competition machine as did almost every other manufacturer (save perhaps George Brough) and such Velocettes as appeared in trials were mostly privately owned by clubmen who no doubt rode them to work as well. The most regular soloists were the Tiffens, father and son, who with great dedication stuck to the Velocette marque when it might have been more expedient and profitable to have switched to a marque which had developed trials machines. In the sidecar lists had been Stuart Waycott, who made his name on a 350 cc iron KSS and rode to ISDT standard by his own efforts rather than by factory influence. While the firm might not have approved of half-way mud plugs, the ISDT was a different matter. It did develop the breed of sports roadsters and with mounting opposition from abroad it was important to British prestige.

Intensely patriotic, as natives by adoption so often are, the Goodmans did their utmost to provide Waycott with the best sidecar outfit they could contrive. As finally developed with a TT engine enlarged to 600 cc, an ingenious sprung chassis developed by Australian Phil Irving and the lines of the Australian Goulding chassis, interchangeable wheels all round and a host of detail refinements, of which the hole cored through the fuel tank to take a special extracting plug spanner was the best known, it was perhaps the most sophisticated British outfit produced. And when, with war clouds gathering over the industry though not perhaps visible to outsiders, the other big makers opted out of the ISDT sidecar class the whole responsibility fell on Velocettes who shouldered it without thought of cost. If ever an outfit ought to have been preserved as a memorial to outstanding achievement by rider and manufacturer alike in international motorcycle sport, it should have been this Velocette. Alas, its present whereabouts are unknown.

ISDT outfit remained at the factory during the War, being used as road transport by Percy Goodman and by Charles Udall, but then it was sold and all trace lost. An unconfirmed rumour is that it was last seen as a grass-track outfit in the West Country. There may still be a chance that it will turn up and can be restored to its ISDT trim. Nothing, I am convinced, is impossible when it comes to famous machines. Two events which have occurred recently convince me of this and make my researches well worth while. I am happy to report, for instance, that "Whiffling Clara", the one-off Willis experiment in forced induction I discussed in an earlier chapter, is still in the land of the living.

I have actually seen her and am satisfied that, although years have taken their toll and modifications intended to turn her from a temperamental prima donna to a maid of all work have disguised her heavily, the bare bones are hers and all being well she may race again. I doubt, though, whether she will whiffle again, as to reinstall a supercharger would mean too many surgical operations. I found, too, thanks to the perspicacity of a northern enthusiast, that there was indeed a supercharger bearing the name Foxwell, the product of a firm called Daniel Foxwell, and it was a development of their established range of industrial vacuum cleaners, high-power rotary-vane air pump used for sucking dust from factories and cinemas, etc. It makes me wonder if the legend of Willis first supercharging an engine by jury rigging a vacuum cleaner was a bit misleading. We all jumped to the conclusion that it was a domestic cleaner!

The other event which proves that the most unexpected relics sometime turn up under one's nose provided me with a memory test and eventually a flash-back to the 1938 TT races. The race I will deal with later, but the memory test began when a local enthusiast rang me for information about a Velocette he had acquired. He thought at this stage that it was a Mk. 8, although there were some peculiarities about it. In conversation he mentioned that the girder forks had an unusual damper system the like of which he had not seen before. The mention of something unusual was enough to arouse my curiosity and I went to see it. The forks *were* unusual. Instead of the normal friction damper on each lower fork spindle there was a circular housing on either side about the size of a friction damper, but it was not concentric with the fork spindle and was actually part of the lower fork link. Or, to be more accurate, the circular housings had lugs or extensions diametrically disposed so that they took the place of the usual fork links. No friction discs were visible. The housings were bridged by a tubular member like a secondary fork spindle housing. This was slotted near the offside to take a wedge to which was attached a threaded rod reaching upwards through a lug on the front of the top spindle bridge and topped with a knurled alloy rod. The function of this adjuster was obvious. When the knob was turned, it pulled the wedge through the slot, and forced apart two plungers which acted on whatever was inside the circular housings.

Something about the shape of the housings, slightly hollowed out on the outside face, rang a bell. It reminded me of an experimental springing arrangement devised by someone named Deane in the late 30s. There was a spring fork

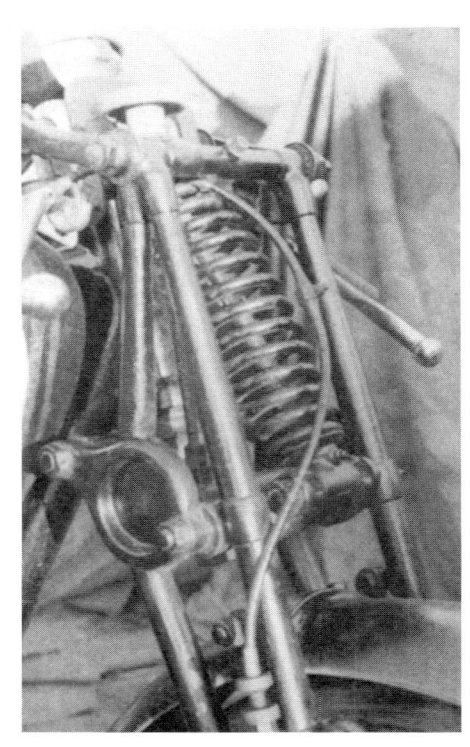

The "mystery" front fork on a pre-war racing Velocette, believed to be the experimental Deane Suspension used by M.D. Whitworth, who was sixth in the 1938 Junior TT

and a spring hub both worked on the principle of large ball bearings rolling in circular cup-shaped depressions against the pressure of compression springs. The fork had a ball assembly each side.

The spring hub had a number of ball assemblies - half a dozen perhaps - in a hub about the size of a Triumph sprung hub. It was, I would think, doomed to failure because to provide any spring movement the wheel spindle had to run eccentric to the spindle. The mechanism would therefore have to operate on every revolution of the wheel when loaded... wearing itself out and wasting power in the same way as running on a soft tyre. The clever thing about the Triumph sprung hub was that the wheel remained concentric to its real centre, which was the inner hub, which was then sprung in relation to the rear wheel spindle. It was in effect a plunger spring frame with the plunger inside the hub and there was no mechanical loss caused by rotation. The fork, however, had possibilities and I remembered that it was tried out at Brooklands by Hugh Trevor-Battye, who was a regular Brooklands habitué, often on a Scott.

So I did a little browsing through the files, turned up the

Billy Tiffen Jnr, after evening practice for the 1939 TT on his Mk. VIII KTT.

first description of the Deane suspension and a later announcement that it was to be produced by the Central Wheel Company of Birmingham. The next clue turned up in reports of practice for the 1939 Junior TT. M.D. Whitworth, it was announced, was to ride a Velocette entered by Trevor-Battye and fitted with an experimental fork damping mechanism - now called the Deane Superior Suspension system. In the race Whitworth held sixth place apart from the second, when he was displaced by Harold Daniell, and was the first true privateer home. His mount was one of the new Mk. 7 KTT models which had been announced shortly before the TT, but must have been available to those "in the know" some time before. The Mk. 7, as I have suggested before, can be best described as being a rigid Mk. 8. The frame was not unlike the production KSS and MSS, but was shorter in the wheelbase.

In the light of experience with weight distribution, the front down tube was at a steeper angle and the engine further forward. The girder forks were exactly the same as the works Mk. 8 type models, the spring having Silentbloc mountings and a telescopic guide. The brakes, however, were the Mk. 5 type with normal hubs, not the conical elektron hubs. On close examination, my friend's mystery model turned out to have a "home-made" spring frame with box-section rear forks instead of the genuine tapered pattern and it would seem that the rear sub-frame was a workmanlike adaptation of a rigid frame. The story was that the machine had once been ridden by Geoff Monty, and as the frame looked very much like the work of his one-time partner, Dudley Ward, it fitted. My guess is the cycle parts of this machine are those of the machine which Whitworth rode so brilliantly to sixth place in the 1938 Junior aided by the Deane Superior Suspension which appears to have acted as a form of check spring to damp the action of the main spring, with the advantage of being readily adjustable by the rider whilst in motion. The engine proved to be a later Mk. 8 type, possibly fitted in the 40s.

The machine has changed hands since I saw it and I hope that the new owner converted it back to its original rigid form, as a memorial to the inventive genius of Mr. Deane and the gallant effort by a well-loved rider, the late David Whitworth.

It can be truly said that the intense rivalry between Velocettes and Nortons in the 30s sustained interest in the Junior and Senior TT races. The fact that it had, in the middle years, become a little one-sided, with the Hall Green Davids squaring up to the Bracebridge Street Goliaths, won the crowd's admiration. But no one likes to cheer a loser all the time and it was fortunate for the sake of the races that Velocettes, with the help of Stanley Woods, began to achieve parity. The keener students of the game began to feel that it only a matter of time before the black and gold triumphed over the black and silver.

The moment they had been waiting for came in 1938.

In the Junior race Stanley Woods set a pace which none of the Nortons could match and won with such nonchalant ease that the race seemed tame by comparison with previous struggles. Equalling the lap record on his second lap and breaking it by a mere two seconds on his third, he thereafter rolled it back, content to win by four minutes from his second string, Ted Mellors, whose task it was to keep ahead of the Nortons and no more. Racegoers pondered at the time what happened to the Nortons rider who, by comparison, seemed not to be really trying. The reason, it emerged later, was that for once Joe Craig had guessed wrong... had geared high for a calm day and was hopelessly overgeared for the blustery one which emerged.

But even the Norton camp did not begrudge Velocettes their overdue victory. "If we cannot win", said Mr. Mansell, the Norton boss, "I know of no better place to which the Trophy could go. They have chased us for years and certainly deserved their success."

This was the truly sporting spirit of the Tourist Trophy races in the pre-war years. The Velocette victory was not quite according to form because their machines were virtually indistinguishable from those of the previous year while the Nortons were quite new. Frames and engines had been strengthened yet lightened by extensive use of light alloy, and the imposing machines were fitted with the then "new fangled" telescopic forks. Joe Craig had always studied the opposition closely... he had taken due note of the speed of the abortive dohc Velo of 1936 and developed his own version of it, and he had obviously been impressed with the action of the tele forks on the 1937 BMW which Jock West had piloted into sixth place in 1937. The Norton fork lacked the hydraulic damping of the BMW — the forerunner of all our post-war tele forks — but from the point of road-holding and, particularly, front wheel braking, it was superior to Norton's girder fork. (The riders apparently had taken some convincing of this until comparative tests at Donington had proved the point.)

Junior practice suggested the Velocettes might have the edge, for Stanley Woods twice lapped at 83.4 mph and Mellors twice at 81.1 and once at 82.6, but the best Norton speed was Daniell's 83.1, with the rest of the team below 80 mph Even in the heat of the chase the race none of them could do much over 82 mph

The real interest in the race lay in the tussle for the places. Les Archer, on a rigid Mk. 7 KTT, at first spoiled the Norton formation by separating Daniell from White and Frith until his engine failed and even then David Whitworth, riding a wonderful race for a newcomer fresh from *The Motor Cycle* Clubman's meeting at Brooklands and on the privately entered Mk. 7 with the Deane Suspension forks, kept the Norton private owners at bay. The winning Velocettes finished in perfect trim and as clean as new pins. The fully enclosed valve gear was really vindicated.

If the Clerk of the Weather had taken a hand in helping Velocettes to their first victory since 1929 he turned the tables in the Senior. This time Velocettes guessed wrong and were slightly undergeared, with the result that after leading on laps three, four and five Stanley Woods was beaten by 15 seconds after a fantastic last-lap spurt by Harold Daniell at a record 91 mph, a speed which was not approached again until 1950 (Geoff Duke, Norton, 93.33 mph).

All in all, the 1938 Senior was just about the closest, most heart-stopping tussle between Velocette and Norton — more truthfully, between Stanley Woods and his Velocette and the full Norton team of top-liners. Stanley was unquestionably the only man who could have taken on the Norton team single-handed. Long-distance races are not won by riding flat-out all the way. They are won by going no faster than is necessary to win, though often this means going fast enough to blow up the opposition and involves the risk of blowing up one's own motor. When you have a team of three star riders at your disposal, as did Joe Craig, you can play the game in various ways. Send one out to set a scorching pace, perhaps sacrifice his chance of finishing to break up the opposition. Or juggle with your men, urging first one and then another forward, to confuse the opposition who do not then know from which direction the real danger is coming. In fact, I was told by Daniell himself that Joe Craig never actually divulged his real race plan to his riders but secretly arranged subtle differences to their machines. The man who Joe had marked down to lead the team would therefore find, when given the requisite pit signal, that he had the legs of the rest. Stanley Woods, with his private signalling stations at Sulby and on the Glencrutchery Road, was the only man in a position to counter Joe Craig's plans.

But one man, no matter how brilliant, can seldom beat an organised team and then only if he has speed to spare.

There were no speed traps in those days so we shall never know what the differential was between Velocette and Norton, but bearing in mind that the Nortons had been extensively redesigned, had the advantage of dohc, and tremendously powerful front brakes made possible by the telescopic forks (the elimination of brake judder inherent with girder forks was rated as the great advantage of the teles), while the Velocette was to all and intents and purposes the same as in 1937, plus a winter's development in the test house, it is reasonable to suggest that the Nortons were faster. If Stanley did slip up in his assessment of the situation it was probably on the first lap which he covered comparatively slowly in 25m 53s, which gave Freddie Frith, the Norton leader, a 25s advantage. But of course Velocettes were known to take a long time to warm up. Frith lost 7s to Stanley on the pit stop at the end of the second lap and, having slowed on the road, lost his lead by 3s. On the third lap Frith was 5s behind Stanley, on the fourth lap he pulled it back to 1s, on the fifth it was back to 3s, but dead-heating with Frith in second place was Daniell, who had started slowly down in fourth place and had got faster and faster. Unnoticed at first by the crowd, perhaps unnoticed by Stanley's signallers. Stanley then equalled the 1937 record of 90.27 mph but Frith retaliated with 90.44, and while they were dead-heated on total time Daniell shot by them both with a new record of 90.75, the first 24m lap ever. So it looked as if Joe's plan was for Frith to keep Stanley busy in the early stages without pushing him too hard, while all the time Daniell was the real danger. At the start of the last lap only 5s separated the three. Stanley put in his characteristic last lap spurt at 90.5 mph, which gained him 2s on Frith, but was of little avail against Daniell's fantastic 91 mph Although Daniell's last lap was a milestone in TT history he always made it seem quite unexciting when questioned about it. No spine-chilling tales of dicing with death.

Just a matter of letting the engine go a bit higher in the gears and braking later. He made it sound so easy. My view is that in all probability Joe Craig had planned it all before the race began.

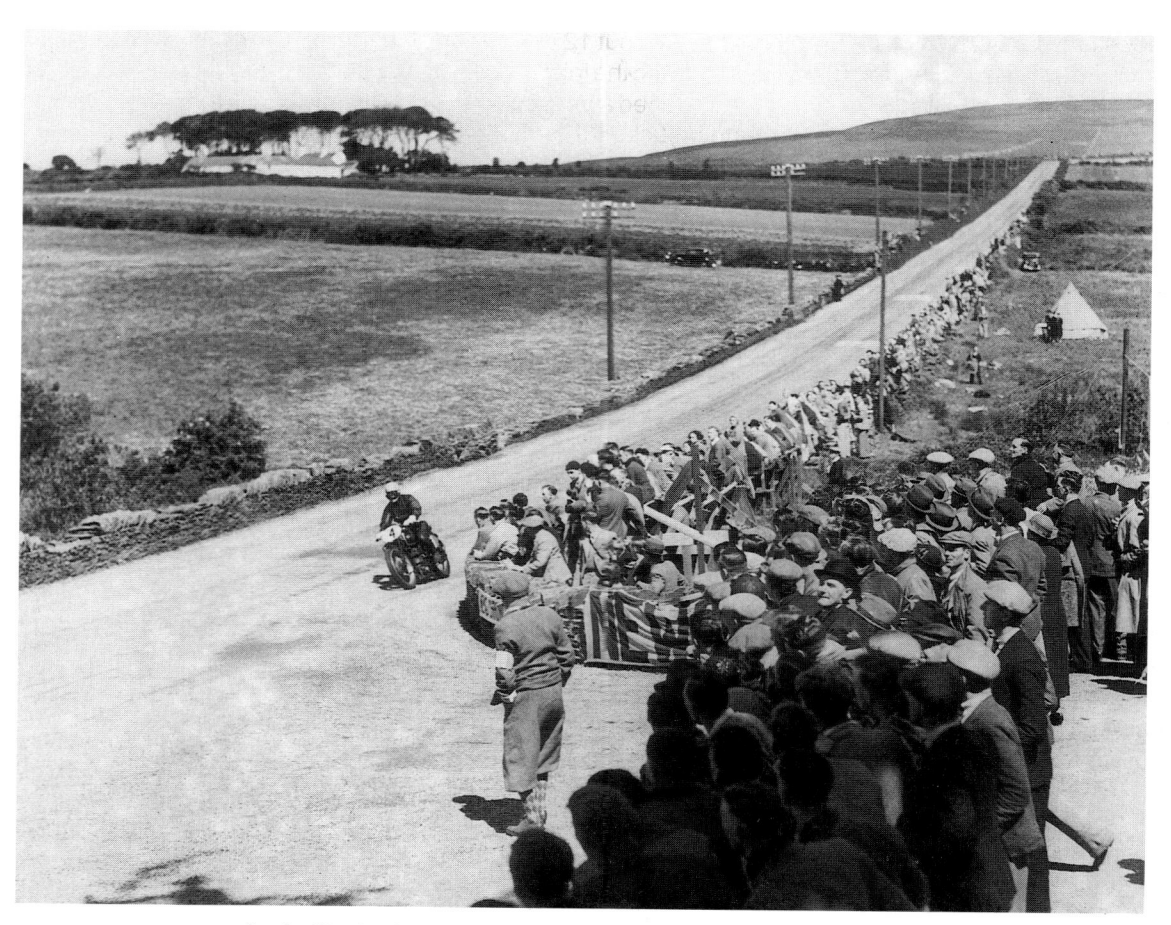

Stanley Woods takes Hillberry, Junior TT, 1938; his first TT win on a Velocette

The Wilkinsons — a 'Velocette family'

YOU MAY think the following mini saga ought to be in the Guinness book of records and I would agree but it serves to show what a lion's heart there was beneath the rather meek appearance of the pre-war MSS.

This Velocette story goes back a long way, back in fact to the late sixties and the early days of Vintage MCC racing when there was such a shortage of sidecar entrants that the class was often in jeopardy. My son Roger was determined that anyone who could handle a sidecar outfit and had anything remotely suitable should have a go and one who was arm twisted into it was Pete Wilkinson who had a Brough Superior, though a rather pedestrian side-valve. His protests that he had no suitable sidecar were silenced by Roger loaning him a chassis and a plank of mine. I was already conscripted in a PR capacity to convey a chap with a cine camera (a Velo man, Dave Ward, actually) who was going to get some action shots of what vintage sidecar racing was all about to encourage (discourage?) others.

I had an old ES2 Norton (it cost £12) to which I clamped a single seater sports body and the intention was that I would enter the sidecar race so as to be on the track (it was Cadwell Park) and try to keep out of the way of the racers. Despite my good intentions I soon got carried away and began to race and the resultant film was hardly documentary, for the passenger became more concerned with his safety as the race went on. For the next meeting I had a plank instead of a sidecar body and I've been racing the same old Norton ever since, though now it's on grass rather than the hard stuff. Pete Wilkinson caught the racing bug too and decided that the Brough lacked acceleration and brakes and cast around for something more suitable. Being a Venom man he had quite a few Velo bits so the obvious bike was a pre-war MSS, the VMCC 25 year rule outlawing Venoms at the time. A far from concours MSS turned up (it had been used for grass track racing) and Pete started tuning based on the advice of one Laurie Nunn, a Cambridge man who had been sprinting one with some success.

It was Nunn who told him that a piston from a Coventry Climax racing car engine (the war time fire pump engine that became a race engine) was ideal to up the compression ratio to about 12 to 1. It was and it did. He bored the inlet port out to 1 in 3/16th and fitted a Venom inlet valve. "Its edge touched the plug hole and everyone said the iron head would crack but the two things that survived on the two engines we have used have been the heads. I fitted a remote needle carb, a Venom cam and followers and a Venom oil pump with a spacer collar. It produced a lot of top end power but no bottom end, that's what you want for sidecar work, so I went to Methanol and fitted a TT carb because the RN one wasn't suitable for alcohol. I tried a close ratio gear cluster but that was no good for sidecar work. Brakes? At first I used the standard steel front brake but after I had changed to the cast iron drum I never wanted anything better. It's a very good brake."

In this form the MSS was a very competitive machine in the vintage sidecar class but as the 25 year date line moved on and allowed 650 Triumph twins, for which of course go faster parts were readily available, more power was needed.

"It wasn't so much the maximum speed as the acceleration of the twins that gave them the advantage" says Pete. But I am running ahead of the story and there was another factor to

Rosemary Wilkinson, with daughter Kim hanging it all out

consider. His wife Rosemary, an experienced rider of his Brough outfit on the road, had started racing it and was already asking for more acceleration and better brakes... my son Roger passengered for her in the early days and urged her on. What was obviously needed was another MSS outfit, to make tuning and spares easy. The late John Griffith came up with a part restored one to which went Pete's sidecar (a Watsonian chassis with the 'wobble' wheel mounting welded solid) and in went Pete's well developed 500 cc motor as soon as he had finished his new 'jumbo' 560 cc job with a purpose built light tube racing chassis. Making an oversize post vintage MSS is not that easy because the barrel cannot be bored much before the base spigot is lost.

Pete got over the problem by enlarging the bore to take a thin liner which itself formed the spigot. In went an 86 mm Venom piston which gave a capacity of 560 cc and a compression ratio of about 13½ to 1 with a ¼in compression plate under the barrel. As the head was still suited to an 81 mm bore there was a ready made squish band. Alas, the power was now too much for the drive side crankcase and although he always used the parallel roller main bearing he got through five or six cases before he gave up and dropped the cr to around 11 to 1. "I even tried a steel top hat to house the bearing but it just made a bigger hole when the crankcase gave way".

In that form the big MSS outfit gave him many years of hard racing with reliability, apart from the con rod breaking once, until he was loaned a potent 650 Triumph outfit and eventually launched in Formula 2 on a Yamaha kneeler. A heart attack grounded him in the end and he turned his attention to the administrative side of vintage racing "to put something back into the sport".

Pete's misfortune was Rosemary's good fortune for she took over the 'jumbo' motor (which is why on an occasion at Oulton Park she left my 630 cc Norton outfit for dead and only some drastic tuning between races and some distinctly hairy cornering kept me in front in the next race).

Now this is where the Guinness type record comes in. Believe it or not Rosemary, is as I write getting ready for her 20th season of vintage sidecar racing and only twice has she failed to start; once when the mag failed and once when the oil pump packed up. Careful lady that she is, she always checks that the oil is circulating before she comes to the line. This time it wasn't. A main bearing failed once through old age and a crank pin broke but she still finished. She's had a lot of sidecar passengers, including her daughters Steph and Kim... and my granddaughter Becky. It must be a record.

Pete Wilkinson reminded me that I once built up a mildly tweaked post vintage MSS for solo racing and loaned it to him a few times. Actually one or two friends rode it in Vintage Club races, Steve Woodward in particular, and I only rode it twice, as far as I can

In winter Pete Wilkinson kept fit by riding in Vintage Club trials. On a Velo, of course. This vintage GTP was tuned by two-stroke boffin George Silk (remember the Silk motorcycle?) and has reworked ports and modern crankshaft seals. "Wonderful bottom-end power," says Pete

remember. Frankly I was not happy with it. It was a bit too big and heavy for me and I didn't find the steering gave me confidence. It tended to weave a bit on the straight, apparently because I gripped the bars too tight. No one else complained.

It was rather like the way a horse senses when a rider lacks confidence and will 'try it on'. If I had it now I would try slightly longer top fork links to give more castor. Besides, it was so smooth and quiet running that it was deceptively fast. I seemed to arrive at corners sooner than I intended. The only trouble we had with it was a tendency to 'wet sump' when it got hot because the standard oil pump could not scavenge frothy oil. I should have fitted a Venom pump or at least a new one. I overcame the trouble by reducing the feed a bit, not so dreadful in practice for as long as the oil is going back to the tanks the engine is getting enough.

Rosemary Wilkinson and sometime passenger, daughter Steph

Steve Woodward riding Titch Allen's MSS racer at Cadwell Park, 1979

Veloce horses

AS WITH any company involved in the 'sharp end' of competition, Veloce never disclosed any brake horse power figures for their racing machines. Just before WWII, Frank Mussett, an Australian Velo dealer, purchased a number of machines and spares from the factory on the basis of "That's all we have, take them and don't ever write in for any spares, you have everything". At some point later on, Rod Coleman, former AJS works rider, was given brake horse power readings for the Mk. 8 and the 500 engines that Mussett took back with him. The figures have been checked by Rod and Dave Holmes, and as an aside we have the figures from the LEDAR dyno for their rapid MOV and MAC racers. Interesting to note that the Holmes MAC is now producing more power than the works Mk 8 did forty four years ago!

Tests conducted by Veloce Ltd.
Information supplied by Rod Coleman
Calculations by David Holmes

500 cc engine for Australia,
Sept. 23rd, 1936
Bronze head.
Cast iron barrel,
Inlet valve 1 3/4in.
Solid exhaust valve
Std 500 con-rod
1934 piston = 75cc chamber, 7.6: 1
1 1/8in carb 360 main jet

Actual Motor RPM	BHP un-corr @ crankshaft
4,500	26.2
5,000	30.9
5,500	33.6
6,000	35.6
6,500	35.2

F. Mussett 350 KTT
(possibly standard Mk. 8 on 50-50 or pre-war big-fin works 350)
Combustion 35cc chamber
1 3/32in carb   380 main jet   50-50 petrol/benzole

Actual Motor RPM	BHP un-corr @ crankshaft
6,000	28.4
6,500	28.4
7,000	26.7

F. Mussett 350 KTT
Compression 13: 1
1 3/32in RN carb   1100 main jet
75% methanol / 25% petrol/benzole

Actual Motor RPM	BHP un-corr @ crankshaft
5,000	25.6
5,500	28.8
6,000	31.0
6,500	31.4
7,000	31.3

Tests conducted by Veloce Ltd.
Information supplied by Rod Coleman
Calculations by David Holmes

F. Mussett 350 KTT
Compression 13: 1
1 3/32in TT carb   1100 main jet
75% methanol / 25% petrol/benzole

Actual Motor RPM	BHP un-corr @ crankshaft
5,000	26.0
5,500	29.0
6,000	31.2
6,500	31.4
7,000	30.3

Engine 5004, May 27th, 1939
New valves @ 1-15/16in
1-23/32in ex (Big valves)
New crankcase, piston & rings, etc.
54.5 cc chamber
1-3/16in RN carb 680 main jet

Actual Motor RPM	BHP un-corr @ crankshaft
5,000	32.9
5,500	36.2
6,000	38.0
6,250	38.0
6,500	37.3

Comment by David Holmes: "Even by 1939 they must have had the combustion fairly good to run at 30° mag timing!"

Engine 5005, May 26th, 1939
Small valve cylinder head 1-13/16in
1-23/32in ex
New flywheels, piston rings etc.
52 cc chamber
1-3/16in RN carb 640 main jet

Actual Motor RPM	BHP un-corr @ crankshaft
5,000	35.5
5,500	39.2
6,000	40.6
6,250	40.2
6,500	38.3

Comment on sheet: "Best 500 in 1939 (small valves)"

Tested on the LEDAR chassis dyno giving rear wheel BHP
Calculations by David Holmes

Holmes MAC
Compression 12.5: 1   Ign. 33 deg btdc
1.187in TT carb   1300 main jet
Methanol 100% plain ex. pipe 36in long

Actual Motor RPM	BHP un-corr @ crankshaft
5,000	24.4
5,500	26.7
6,000	28.9
6,500	31.0
7,000	32.8
7,500	34.0
8,000	33.8
8,500	33.1

Holmes MOV
Compression 13: 1   Ign. 28 deg btdc
1.330in TT carb   1400 main jet
Methanol 100%   plain ex. pipe 33in long

Actual Motor RPM	BHP un-corr @ crankshaft
6,500	24.3
7,000	26.1
7,500	27.6
8,000	28.8
8,500	29.8
8,750	30.1
9,000	29.7

Holmes MOV
Compression 13: 1   Ign. 28 deg btdc
1.330in TT carb   1400 main jet
Methanol 100%   megaphone ex. system

Actual Motor RPM	BHP un-corr @ crankshaft
7,000	24.5
7,500	26.9
8,000	29.0
8,500	30.7
8,750	31.2
9,000	31.5
9,250	31.1

Peter Goodman: boss's son who had riding talent

He had two periods near the top of the road-racing tree

Peter Goodman waits for his mechanic to fit the hard plug, in the paddock at Cadwell Park in 1946. Watching thoughtfully is Sam Coupland, the Boston tuner who was the man behind the scenes in many pre-war and post-war Velocette and Guzzi successes. The sprint tank reveals the massive upperworks of the later development of the ohc Velocette; this is a works 10in head sohc 500 model

Around this point in the Velocette history a schoolboy comes on the scene, a schoolboy who must have been the envy of his friends, for he was the son of Eugene Goodman and on his 16th birthday he became the proud possessor of an MOV Velocette. Fell off it on the first day too, he recalls with a reminiscent smile. The boy is now, of course, a man and has a son of his own, and in the way of the Goodman-Velocette dynasty became Works Director, having almost literally stepped into his father's shoes. But between the schoolboy setting forth on his first Velocette and the man responsible for the production side of Veloce Ltd. there were many years and quite a story. The story of Peter Goodman who had two periods near the top of the road racing tree, with a war between them, and might have gone on to even greater heights in racing but for a serious racing crash which was not his fault at all.

His name is not Peter at all. It so happened that when he was christened his father was busy at the works and when his wife told him how she christened the infant he announced: "I shall call him Peter." And so he did and so did

DOX 665: Built specially for teenage Peter Goodman, this MSS had a tuned engine with a bronze head and was capable of 90 mph. Frame is believed to be Mk. 6, forks are braced Webbs and front brake is KTT

everyone else until it became his name by common usage, and he has always used it himself. It caused a little confusion sometimes over official documents like passports, but I do not propose to add to the confusion by saying what his Christian names are. I only mention the matter at all because it illustrates the dedication of the Goodmans to their work.

On leaving school Peter Goodman was apprenticed to Alfred Herbert, the Coventry engineers, and soon began to nag his father for a machine with more "steam". Eugene came up with an intriguing "one-off" which was typical of the specials which emerged from Hall Green at times. It was a warmed up MSS engine with a special bronze head fitted into an ex-racing bike. Peter is pretty sure that the bicycle parts were Mk. 6 KTT. By all accounts this MSS-KTT special was quite a bicycle and with its 90 mph maximum gave Peter Goodman, apprentice, a good grounding in high-speed motorcycling. Especially as at that time Franz Binder, the Continental privateer, was doing a stint at the works putting miles on a hush-hush pivoted fork MSS evolved by Phil Irving, and Peter often went with him on long high-speed proving trips.

Naturally he wanted to try his hand at racing. Donington Park was the place to start as organiser Fred Craner always saw to it that there were novice classes for riders on the bottom rung of the ladder. Peter entered for one of these on his roadster special. When Eugene found out he was not best pleased.

"If you want to start racing you must have a proper bike," he said, and vetoed the idea of racing the special. But he did at once set about providing a proper bike. There was not much difficulty in obtaining a KTT, but the tuning and preparation of it did present a problem for the race shop was far too busy to take on a machine for a beginner. Eugene solved the problem cleverly by persuading the repair shop to take on the job. You can well imagine that this would be

Peter Goodman heaves his KTT into life at Cadwell Park, Whitsun, 1946. No. 4 is George Brown with a TT replica Vincent-HRD

a wonderful morale booster for the chaps working on humdrum roadster repairs and that they would do their damndest to show "those toffee-nosed mechanics in the race shop" that they, too, could tune a bike. Which, of course, is the way it worked and eventually there came a time at Donington when Peter on the 'repair shop' bike blew off the works machines with no less than Stanley Woods and Mellors aboard.

This was because — and Peter, the most modest of men, is quick to point it out his machine had been patently prepared and modified for a short circuit, whereas the works machines were Isle of Man models and not really suitable. There was, for instance, the warming-up problem. With a gallon of oil in quick circulation a KTT needed perhaps 20 miles to give of its best, by which time a Donington race was as good as over. Most people, including the Velo works men, heated the oil over a primus stove and poured it in at the last moment. Eugene did not like this rustic method and evolved a "secret-weapon" which ensured that Peter's bike *went* from the word go. So simple, really. A perforated container about the size and shape of a cocoa tin was built into the oil tank over the outlet pipe and under the return pipe. When the engine was started it drew on the third of a pint of oil it contained and then went on circulating this oil until gradually the main supply percolated through the holes. It was like having but a third of a pint to warm up but a gradual topping of cool oil. I have no doubt there were other crafty ideas emanating from Eugene and much elbow grease exerted by the repair shop lads.

And before Peter came to the starting line his father saw to it that he got a bit of track practice. The works were testing Waycott's ISDT sidecar outfit on the full circuit at Donington and Eugene arranged for Peter to go along and do some lappery. He also arranged for Billy Wing, the Nottingham rider agent and no mean rider at Donington, to show Peter the way round. Years ago, Billy told me, anent this little operation, that after a few laps Peter was showing him the way round and he told Eugene: "I can't teach this lad anything."

This way Peter got in about 100 laps of the circuit before

A photograph from P.J.G.'s personal scrapbook. The Veloce delivery lorry decked out for the Redditch Carnival

Peter Goodman eases the aches and pains of 264 miles of high-speed TT racing with a cigarette. He had just finished third in the 1947 Senior TT, sharing the fastest lap with Artie Bell (Norton)

his first novice race, in which he finished fourth. At his next outing, August, 1938, he won his heat and thereafter Craner put him in the expert races. He won his heat in the April, 1939 meeting — he points out that Craner had put him in the easy heats — but in May he came up against Harold Daniell and beat him by 18 seconds.

I could tell that this was a high spot in his memory but he was quick to add that one reason for his success was that his father had made up a higher than standard second gear and had done some work on the motor.

The race report reads: "P. Goodman rode a magnificent race."

In the unlimited class he won the first heat on his three-fifty but was outclassed in the final results.

At the Dunlop International meeting later he beat the Velocette work's team of Woods and Mellors, which he says must have caused some embarrassment, and won a free entry to the Manx. Alas, there was no Manx G.P. in 1939 and instead Peter Goodman went into the RAF. He was the sole survivor of a plane crash but he does not say much about that, preferring to tell a story against himself of an impromptu wartime scramble with a twist of coincidence in the tail. "Towards the end of the war I was stationed at Darley Moor (now a short circuit) and there was a scramble nearby. Alf Briggs lent me a machine and I was chasing him when I fell off. Who should pick me up but Billy Wing. And he didn't half tell me off for not wearing leathers."

In a brief two-season apprenticeship to big-time road racing Peter Goodman had made his mark at Donington, and the best proof of this is that the "books" started referring to him as Peter Goodman and no longer plain P. Goodman! When they got down to Christian names in those days it meant you had arrived. But there was no mention of him being the son of a director of Velocettes.

Out of the RAF, Peter started work at Hall Green, his first job being to start up production of the KTT model. He rode in the Manx, finishing sixth in the Senior and fell off in the Junior, and did his rounds of the short circuits at home.

His five-hundred was the pre-war ex-Stanley Woods model, the post-war dohc model. The Senior Manx was the never to be forgotten one run in blinding rain and fog and won by Ernie Lyons on the prototype GP Triumph, it being said that he was the only rider who could find his way in the murk. Peter Goodman, a master of understatement where his own exploits were concerned, made no comment about this race. But he did recall something of a "kafuffle" about his eligibility for the Manx owing to his having competed that July, and won a class, in a rather ragtime event called the Grand Prix du Zoute in Belgium. Despite the rather grand-sounding title and an attempt to designate it the Anglo-Belgian GP, it did not in fact have any international status and was no more than a bit of good fun hatched out by the Federation Motocycliste Belgique and our Sunbeam MCC to get the Continental Circus going. The circuit was no more than 1½ miles to the lap, the straight was a mere 400 yards and the average width 16ft, while the surface was variously constructed of asphalt, concrete and manholes thrown in. Few riders got into top gear on the circuit.

It has an important place in the story for Peter Goodman for, apart from the race whetting his appetite for Continental racing, the FIM timekeeper, Monsieur Lamot, accorded him the record lap of 61 mph, achieved when chasing the five hundreds on his three-fifty to finish fifth in the Senior race. Gaston Lamot was later to become his father-in-law. "But I didn't know him when he put me down for the record lap," he adds quickly. Before international road racing got going there was a lot of fun to be had on the short circuits at home — Olivers Mount, Scarborough, Cadwell, Abridge, Anstey — in which Goodman came up against the short-circuit stars like Tommy Wood, who was usually a hundred yards ahead before anyone else had the clutch in. There was Shelsley Walsh hill-climb where he won the 350 cc class and earned the comment "outstanding". An ex-Isle of Man Velocette was not always the ideal machine for short circuits, but he had a successful season — he was usually placed — and like many other racing men in those days, turned to trials in the winter.

Riding a modified Velocette MAF model, the war-time version of the MAC, he won a first-class award in the Kickham and nearly got a ride in the British Experts, but was ruled out over a technical impasse. If the GP du Zoute had been a real international GP he would have been eligible for the Experts, but the fact that his entry had been accepted for the Manx discounted it.

In 1947 road racing was back to normal with the Isle of Man TT and the Continental grands prix. Goodman was fourth in the Junior TT despite running out of petrol and having to coast down the Mountain on the last lap, and was a brilliant third in the Senior only half a minute behind Artie Bell, with whom he tied for the fastest lap at 84.07 mph This was the first time he had really come up against the might of the Nortons in the Island and of this race he recalls: "I really lost the Senior on the first lap. I was number 69 and Harold Daniell (the winner) was number 72. He caught me a lot quicker than I expected and had Ted Frend with him. When I saw Harold in the pits filling up on the seventh lap I thought I had a chance, but he passed me again by Ramsey."

"Was the Norton faster than the Velo?" "Oh, no. The Velo was faster up the Mountain and I caught him again. He was just faster through the twisty bits. He told me he nearly hit me from behind when I slowed for Windy." Peter Goodman is no line-shooter.

He scored his first classic win the next month by leading the 350 cc class of the Dutch TT from the start, but fell off on wet tar in the 500 cc class and hurt his arm, which put him out of the Belgian race a week later. With race machinery at a premium there were many top-flight riders ready and willing to take over his machine and he had intended to lend it to Bob Foster, but a telegram arrived from his father saying that under no circumstances was he to lend his

Moment of victory. P.J.G. wins the 350 race at Strasbourg, then goes out for the 500 — and fate strikes. He will never race again

machine. Not knowing the reason for this pre-emptory instructions, he had perforce to obey. Afterwards he learned that his father, realizing that he would be badgered to death by riders trying to borrow his machine, had sent the telegram to protect him. Of course Bob Foster could have had it. It was the only time that the close understanding between father and son misfired.

"My father was very good to me," he says of his racing years.

And then a month later fate played a cruel hand.

He entered for a meeting at Strasbourg and it was to be followed by a visit to Brussels to see his fiancée, the daughter of the timekeeper at la Zoute. It was not a very important meeting and not a very good course — set tar and tramlines caused a crop of accidents — but Peter won the 350 cc race and started in the 500. He remembers starting but that is all. Somehow a mineral water bottle rolled on to the course. An official, the organizer of the meeting actually, dashed out to retrieve it... straight into the path of Goodman's Velocette. Peter sustained a fractured skull.

So grave was his condition in hospital that he could not be moved by road. Lord Selsdon, a racing car driver with air line connections, arranged for him to be flown back to England. His fiancée accompanied him. It was the end of his racing career. By the time he had recovered his cousin, Bertie, had come out of the forces and had taken over the racing side and Peter began to understudy his father on the production side against the day when he would retire. Like several other retired racing men he turned his interest to boats. First a cabin cruiser, then a converted lifeboat — appropriately the ex-Douglas Isle of Man lifeboat — and now a rather nice Fairey Atlanta.

He does not think much of modern racing from the spectator angle. "They remind me of boiled eggs. With fairings they all look the same."

But in the Goodman family tradition he frequently rode to work on a Velocette. Usually an LE Vogue. Not quite a standard one, of course - it's got a 250 cc engine.

C.G.H. Dunham, son of a Dunstable Velo dealer and friend of the Goodman family, comes a cropper at Quarter Bridge in practice for the 1946 Manx GP. He was riding Peter's bike!

The racing stories of Velo rider Ernie Thomas

THE DIFFICULTY of making a story like this come to life is that so many of the characters who played the leading parts are dead, out of reach, or just cannot remember much about a part of their life which they put aside when they hung up their leathers for the last time. Facts and figures and contemporary reports, so carefully written as not to divulge any information the manufacturer had not authorized, are poor stuff from which to write a saga. It needs the leavening of personal reminiscences and the odd items of information that can only come from someone who was there at the time, and preferably on the inside.

Fortunately when I came to write this section of the Velocette Saga some years ago there was one man very much alive who was "on the inside" from 1928 to 1939... and again in 1949... having ridden Velocettes 12 times in the TT and countless times abroad (I think his must be a record among Velocette riders); and he was Ernie Thomas. As he always had a reputation for being a rare wit and raconteur I was delighted when he agreed to see what he could remember. He *was* a wit and a raconteur, with a better than average memory, as I soon found at the motorcycle showrooms in Stonehouse, Glos., which bore the name Williams and Thomas; the Williams part was Jack Williams of Cheltenham, a consistent TT rider before the war on Rudges and Nortons and an ace trials and scrambles rider.

The story of Ernie Thomas's association with Velocettes was very much the story of his racing life for apart from minor diversions with Triumphs, AJSs, DKWs and a single ride with a lightweight Guzzi, his name became synonymous with Velocette. And it all started with a roadster "cammy" way back in 1926 in the same way that many other riders, some of them to become famous, with perhaps Freddie Frith emerging as the most famous, got their feet on the first rung of the ladder through the all-round excellence of Percy Goodman's inspired design.

Ernie Thomas remembered how impressed he was with the smoothness and mechanical quietness of that first roadster Velocette, and remembered too with pride that in those early days he could cope with the timing of the bevel gears and adjust the clutch. These two operations were in those days, and long after, a mystic rite which sorted out the real mechanics from the ham fists. He was working in a garage at Bristol at the time and because of his enthusiasm the garage sold a lot of Velocettes, which brought him into contact with the factory. He started riding in sprints and hill climbs... Gopsall Park was one sprint he remembered for it was there that he met stars like Handley (and I remember it too for it was the first sprint I ever saw)

On one occasion clutch slip as-

Ernie Thomas - victorious in Germany

sailed him and he was beaten by an AJS. He mentioned this to Percy Goodman, who said "You had better have one of our new seven-plate clutches. We can't have you beaten by an AJS".

He graduated to Syston Park short (very) circuit road races and had some success, enough to impress Velocettes, who invited him to go over to the TT with them. Not as a rider but as a helper, and so it was that he was at the Velocette HQ, the Nursery at Onchan, on the day before the weigh-in in 1928 and was witness to one of those little behind-the-scenes cameos which you don't find in the motorcycle papers.

It seems to have been a rather light-hearted party for they had all been trying their hand at lifting a 56 lb weight. Percy Goodman could lift it with either hand and so could Ernie. Alec Bennett could with one hand but Willis couldn't lift it at all. Larking over, they fell to discussing chances in the TT, and Alec Bennett said he could do with more acceleration. (A valid point. The vintage overhead-camshaft Velocette lacks bottom-end torque as compared with contemporary engines like the AJS, Blackburne and JAP. It has to be kept "on the boil", as they say, which was not an easy thing to do with a three-speed box.)

"Give me a hacksaw", said Percy Goodman. Thomas reached him one down and Goodman promptly cut about a foot from the end of Bennett's exhaust pipe. There was plenty to go at for on those early models the pipe stuck out well behind the rear wheel.

"Try that", said Percy. Bennett tried it up a rough track, so rough that Thomas failed to see how he could assess the result, and pronounced it better. With that the bikes were weighed in. After the race, the 1928 Junior which Bennett won, with Willis second, Willis came out with a typical Willis remark.

Ernie Thomas on the works dohc engined spring frame on which he made a brilliant effort — "his finest hour" — after Stanley Woods had retired early in the 1936 Junior TT. Although attention was focused at the time on the double knocker that let Stanley down there were in fact two other dohc works bikes in the race. Mellors with a rigid frame finished third and Thomas on a springer fourth after a fall. It shows how little reliance you can place on contemporary reports.

"Now if you had done that to my bike instead of Bennett's I could have won and it would have saved the firm a lot of money." The point being that Bennett, as a professional, would have to be given a large share of the spoils — Willis being a member of the firm, wouldn't. The kind of remark that Willis could make with such a straight face that no one could be sure whether he was joking or not.

Next year Thomas was invited to go over as a reserve rider and when Willis fell off on the third lap of the Junior Thomas was given his bike to ride in the Senior. He finished 14th, which was a good effort for a first timer on a Junior machine. The interesting thing is that he says the bike had the one-off long-stroke 68 x 96 mm Willis experimental engine... interesting because, according to the history books, this engine was not used until 1930 when Willis got up to third place with it before running into trouble. It may be that it was on the secret list in 1929. Thomas recalls that it was no faster than the normal engine but had better pulling power. (Mechanic Sammy Jones said it had the stroke of a bicycle pump.)

In 1931 Thomas was 12th in the Junior, in 1932 he retired, but in 1933 was 10th. In 1934 he switched to a work's Senior Triumph, the massive Mk. series two-valve model based on the new Val Page-designed roadsters. The engine gave a lot of trouble, so much that he doubted whether he would get in a complete lap. Not unnaturally, he was subjected to a lot of chaff from his Velocette friends.

"Did you get over the 'hump? (the Mountain) today?" Willis asked. To which Thomas retorted, "There's one thing, it steers better than a Velocette".

Thomas had arrived at a moment of truth. The much-vaunted Velocette steering was not all that hot. There was a fast righthander after Union Mills that he couldn't take flat out on the Velo, because the back wheel hopped out, but he could take it flat out on the Triumph, which was as fast. He began to realize that there was something wrong with the Velocette back end.

"The trouble was that the corners wouldn't keep still" was a typical Thomas way of putting it. His conviction, based on his experience with the Triumph, which let him down in the race when the valves tangled, must have been noted by Velocettes for soon after he was taken on at the works to help with testing racing machines and later he was one of the riders who attended the secret test session laid on for Stanley Woods. The session which established, with the aid of a block of lead the size of a brick clipped in front of the engine, that the centre of gravity needed to be brought forward, not backward as Willis maintained. They experimented with the old "non-steering" frame, as Thomas calls it (it would be a Mk. 4 or 5 type), a new Mk. 7 type and the new springer.

At first, he says, Stanley preferred the rigid frame because of the light weight but, reminded by Thomas of the bumpy bits of the course where the rigid frame would be at a disadvantage, finally chose the springer.

Willis seems to have had a radical viewpoint about spring frames. "I told him that I had followed Tenni on the sprung Guzzi through Glen Helen, and although his back wheel had been jumping up and down he didn't. Willis said, 'You riders want armchairs'. I said that if he felt like that we could have rigid front forks".

As a rider and race tester Thomas made other valuable contributions to Velocette development. There was the matter of rev-counters. During carburettor tests he was asked repeatedly if different settings were making an improvement. He couldn't really tell but had a brainwave "Why don't you give me a rev-counter and then I could tell you what is going on?" he suggested. A rev-counter was taken from the test bench and fitted. "Why don't you leave it on for the race and then I could see when to change up?" was his next idea. It was tried for the North-West 200 and raised the question of which was best, speed wise, 7,200 in third or 6,800 in top. "Udall did some sums and after that I took it to 7,200 and then changed up."

The special fork spring arrangement seen on the Mk. 7 and Mk. 8 models where the spring is pivot-mounted at either end and relieved of side stresses by a telescopic guide was the direct result of Thomas's misfortunes. He was put out of the 1936 Senior by a broken fork spring and the next week in the Dutch he limped home third on the rubber buffers with another spring gone. Back at the factory there was an inquest.

Thomas thought the failures were due to vibration. "You can see two springs sometimes", he told them. Udall did not think vibration was the trouble. Taking a long iron bar, he inserted it between the fork cross members and heaved the forks to full extension. It could be seen that the spring, being rigidly attached at either end, had to bend as it stretched or compressed. So did the springs of every other conventional girder fork. With the redesigned fork the spring had a straight forward job to do and there was no more trouble.

Ernie Thomas with a Mk 6 KTT, Isle of Man, 1937

Another bit of valuable information Thomas contributes to the history of the racing Velocette, concerns the origin of the oleo pneumatic "legs" which made the Mk. 8 type springer so superior to plunger or friction-damped frames.

"Willis was a keen amateur flyer and we often used to hop over to a flying club near Leicester. One day we were watching a plane taxi-ing over the rough field and noticing the way the undercarriage rose and fell. Willis said 'You know, that's what we ought to have for our back wheel'."

Dowty, the pioneers of oleo pneumatic aircraft undercarriage struts, were approached and produced some special units suitable for the spring frame. Air pressure in them was very critical. It was found that when testing with an ordinary tyre pressure gauge enough air escaped to upset the reading. For the work's machines a special gauge was evolved with a stop cock arrangement which prevented the slight loss.

And now it can be told, the story of what I have described as Thomas's "finest hour", the occasion when in the 1936 Junior Stanley went out on the first lap with mechanical trouble and Velocette hopes depended on Ted Mellors and Ernie Thomas. Thomas reacted brilliantly to the challenge and for four laps lay third, a mere 20 seconds behind Frith and 53 seconds behind Guthrie. Then he fell off, lost time and his place to Mellors.

"It was a tricky day" he recalls. "In places the roads were dry and in other places they were damp. There was a wet patch at Quarter Bridge and I came off. I kicked things straight as best as I could and got going, thinking on the way that there would be some nice big stones in the walls out towards Union Mills. With the aid of one of them I straightened things a bit more but they black flagged me and insisted on examining the machine, and that cost me more time."

"The bike made a noise like a dustbin rolling down the road yet I still heard amidst it the tinkle of glass from the broken rev-counter."

After the 1937 series, in which he finished third in the Lightweight on the DKW and sixth in the Junior on a Velocette, he left Velocettes to go professional on the Continental "Circus".

The DKW ride had come about in an odd way. It seems that DKWs, looking for a British rider to add to their team of Kluge and Wünsche, had approached Eric Fernihough. This showed a remarkable lack of Island knowledge for although Fernihough was a record-breaker extraordinaire he was no road racer, as he would have been the first to admit. His only experience of the Island was a retirement in the 1927 Lightweight on a New Imperial and ever after that he had concentrated on track work with the occasional foreign excursion on a lightweight. Ernie Thomas heard that DKWs were stuck for a rider and just managed to get a letter to their emissary, Jack Woodhouse, before he returned to Germany empty handed. (He gave the letter to a passenger train guard at Birmingham so that it was posted in London in time.) He remembers the DKW as a very good machine with lots of power out of the corners though it did seize up on him at Donington and fetch him off.

To ensure rides in the 350 cc classes on the Continent Thomas bought a Velocette from the works before he left and surprise, surprise — for me, at any rate — it was none other than the legendary one and only Mk. 6... the Willis confection of racing bottom half topped by roadster Mk. 2 KSS cylinder head. It had been ridden on the Continent by Mellors in 1936 and Austin Munks had won the Manx on it, and it served Thomas well that season apart from a tendency to break its coil valve springs. Unfortunately Thomas could not for the life of him remember what he did with it afterwards or find a photograph of it so the "What happened to it?" question remained unsolved. It could have gone back to the works to provide the basis of Peter Goodman's MSS KTT.

His last pre-war ride in the Island on a Velocette was in 1939 when he finished 15th in the Senior; he rode once more after the war, in 1949, when he was right down the list in 34th place.

"I should have been fourth but I oiled a plug at Governor's Bridge and the marshal would not let me push start until everyone had gone by. I ought to have claimed the time for the delay but I didn't think about it at the time.

There's one distinction Ernie Thomas claimed for his racing life without fear of contradiction at the time. He was the only rider shipwrecked on the way to a race. It happened in 1938 when he was setting off for a continental race with an AJS. In those days riders, even quite high-ranking ones, had to take their bikes with them by boat, rail and sometimes hack them by road (he recalls riding pukka racing Velocettes to the port. "It ran them in nicely" he says). This time he set sail from London on a boat for Hamburg and got his head down in his bunk. He had just dropped off when there was a bump which threw him on to the cabin floor. The ship had rammed another in fog, been holed and "all hands on deck" was the order. They didn't have to take to the lifeboats for they were still in the Thames Estuary and the skipper managed to beach the ship off Dagenham. Ernie and the other passengers were taken off in a police cutter but it was a couple of days before he recovered the AJS.

One final memory about the "dog kennel" 500 cc racers, pertinent to the story of how Stanley Woods lost the 1936 Senior by 18 seconds because the Velocette developed a misfire at the crucial moment when he was putting in his last-lap spurt.

"Yes", said Thomas "we had quite a bit of bother with misfiring on the 500 and never found the cause. I remember that when Stanley Woods was on his honeymoon in Spain he wanted to ride in the Swiss GP at Berne and they sent me over to look after the bike because they didn't figure he would feel much like mechanicing. He tried the bike out and complained of a misfire. 'Put another magneto on', he said. Not likely, I thought... not on a Velo where you couldn't get at the base nuts and had to play about with shims to get the height right. I dropped the needle a notch and it was all right."

Knowing nothing at all about it, I should not be surprised if the incipient misfire was caused by plug fouling through running on a richish mixture. The 500 model was "near the bone" from a heat point of view, demanded forged pistons, and was finally given a cooling spray of oil to the underside of the piston. This was arranged by running a small-bore steel pipe from the big end up the side of the connecting rod and into the small end. Two small-end bushes were pressed in from either side to leave an annular gap. Oil filled this gap and then escaped through a small hole in the top of the little-end eye, a small hole to give a jet where normally there would be a large radiused hole to collect oil for the little end. An idea used on aircraft and racing car engines.

By a coincidence, on the same day that Thomas told me about this arrangement, I called on Velocette enthusiast Michael Tomkinson and he produced from his collection of unidentified objects, relics of the Hall Green racing department, one of those connecting rods. It can be seen that the oil fed up the pipe on the rod was that which would normally have escaped between the rod and the flywheels. This clearance was kept to the minimum (Thomas remembered it as 4 thou.), the escape grooves on the flywheel omitted and much of the oil persuaded to enter holes in the cheek of the big-end eye communicating with the oil pipe. The pipe was clipped to the web of the rod by a series of tiny saddle clips screwed to it and soldered over.

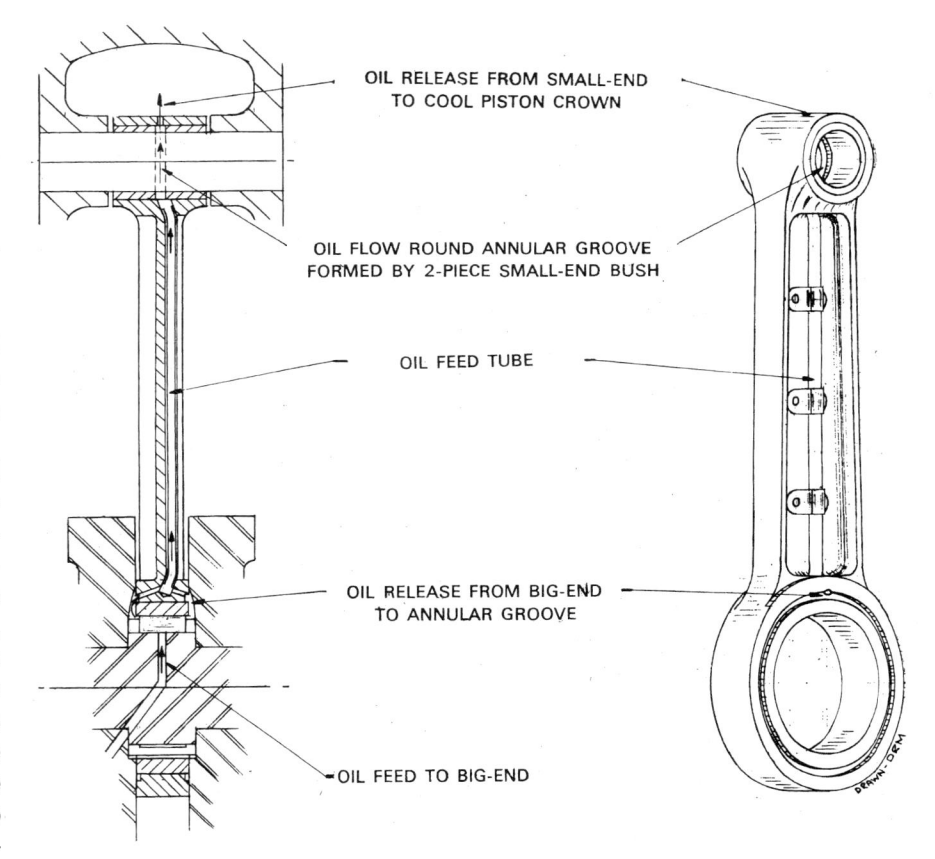

Oil cooling arrangements on the 1936 works 500

Model O
The secret superbike

Whenever I let someone have a brief ride on the O they all say "Isn't it smooth" when they come back. Usually before they get off and while it idles quietly like a car. I thought Peter Goodman would have remembered it well enough from the days when he used to borrow it (when cousin Bertie wasn't using it) but sure enough when he had a ride on the drive at Stanford Hall his first words were still "Isn't it smooth". And so said Alf Briggs, which, as he was a long time technical man with Honda, who had ridden most of their sophisticated multis, I take it to be great praise.

Love at first ride. That was how it was for me when my late friend John Griffith handed me the key to the Model O on an Isle of Man holiday and said, "Use it as your own... you can go home on it afterwards." Within miles I had fallen for its charm and now, after many years and some 4,000 miles, I am still captivated by it. It's so right — for me, anyway. Seat height is low, so I can plant both feet on the ground, the centre of gravity is exceptionally low, so it doesn't try and fall on you, starting is so easy you have to learn to stroke the pedal rather than kick at it. Low compression, auto advance and coil ignition, plus a favourable kick start to engine ratio, make it easier than the average twin. And because the thrust goes by the nearside crank, and not through the gearbox, there is the up to date advantage that you can kick start when in gear by lifting the clutch.

But what is it like to ride? Well I've said before and say again that it is very much like an Ariel Four, even to the *clickety-click* of the coupling gears. It doesn't bark like a twin, it purrs or drones like a four. More torque and flywheel effect than a modern multi. The steering is unshakeable. Punch the bars and it takes no notice. Bob Currie, doyen of road-testers, wrote of it: "It steers with the inevitably of a county council road roller." He didn't mean that it was heavy, just that it was unshakeable by pot holes, ruts or anything normally encountered.

All I can say about the rear suspension is that you are not aware of it, or of pot holes. It doesn't wear itself out dancing up and down to show you that it works. Now the big question. Would it have sold in 1940? Only to mature enthusiasts, I fear. I was not so mature in 1940 and would have opted for a Triumph Speed Twin and saved the extra money an O would have cost — say, 20 percent. I would have preferred the sporty image (and the name Speed Twin)

Works picture of the Model O, showing covers over suspension units (gone missing) and silver painted 'bonnet' panels

to this very sophisticated luxury super tourer.

If of course the Roarer had been a success, had trounced the BMW in the 1939 Senior TT, the image of the O would have been better though it would have needed a KSS-type face lift to counter the staid KTS look, to compete with the Tiger 100. Narrow front mudguard, 21in wheel, up-tilted mega silencers maybe (less baffles too!)

By 1955 I was more mature, covered up to 40,000 miles a year, was fed up with vertical twin vibro and was more concerned with function than image (I actually switched to an Ariel Four). I would have bought the O which by then would have had tele forks, 12v electrics and perhaps an optional electric starter. It could have been a big seller in the upper end of the market, ousting the Ariel Four and the R69S BMW, and it could have made a profit because it was a production engineer's delight rather than the time-consuming collection of bits and pieces in the average bike of the time. The engine is like a car unit, comes out in minutes, as a lump with gearbox and electrics. Apart from the gearbox (an adapted MOV) there is no ball or roller bearings, just plain bearings with pressure lubrication, like a car.

The remarkable thing is that this first and only prototype was so right that it has been hacked and thrashed by many riders, several of them racing men, has done at least 20,000 miles, and is still going strong. The faults have been trifling and the only blow-up due to rider error (no oil in the sump). I have had prototypes that have been rehashed and modified and run into several versions but none as right from the start as this. It is a lasting tribute to Phil Irving, a very down to earth, practical motorcycle enthusiast with a gift for coming up with novel ideas which worked. A tribute too to the Goodmans who gave him the opportunity to try out his novel ideas, being of course practical motorcycle enthusiasts themselves.

As Velocettes in the days of the elder Goodmans were great believers in using racing to develop the breed and promote sales of roadsters, the production of a machine as revolutionary as the Roarer may seem out of character but in fact they did intend to produce a road-going twin embodying the same basic principles, and this was the Model O. It had the same layout of contra-rotating crankshafts — though for reasons of convenience in the use of gearbox internals from the M range, they rotated in the opposite direction to those of the Roarer — and the final-drive shaft was on the offside. Where on one machine there was a supercharger, on the humble cousin there was a dynamo, and while one derived its basic cylinder-head design from the racing KTT the other inclined towards the bread-and-butter push-rod range. The Model O was the work of Phil Irving and it incorporated the adjustable spring frame which remained a Velocette feature. It did in fact have a stressed pressed-steel rear mudguard unit, as on the experimental MSS, which was a daring step forward. Velocettes seemed to have "gone off" the dual seat for the Model O had a saddle and a rectangular tray to take a pillion seat, exactly like the LE.

Bob Burgess recalls that the Model O had a one-pieced light-alloy head on a monobloc iron cylinder casting and that the pushrods came up a hole cored centrally. The rockers were spread out in X-formation, the closely grouped push-rods being in the centre of the X. Duralumin connecting rods ran directly on the crankpins. Coil ignition was

On his last visit to England, Phil Irving is reunited with his brainchild, the Model O. Ted Davis, former Vincent development engineer and colleague, watches while C.E.A. makes an adjustment

used. He recalled too, that the steering was at first not up to Velocette standards but was further improved by lengthening the front forks by cutting and welding in an extra section.

Billy Wing of Nottingham, the one-time Velocette star privateer at Donington, and life-long Velocette agent, was one of the few people outside the factory who had a ride on this machine. He told me that it was very smooth, handled well and would do about 100 mph He thought it was streets ahead of the contemporary Triumph twin but then again he was a dedicated Velocette enthusiast.

John P. Griffith bought the Model O (with engine seized) and exhibited it at Stanford Hall Museum. Subsequently the O was rebuilt by Michael Postlethwaite, and John Griffith used it on the road until it was acquired by me. It took part in the cavalcade of British machines to Brussels when Britain entered the Common Market and has been frequently ridden and demonstrated since. It is usually on view in the Stanford Hall Museum.

Through ownership of the Model O I came to know Phil Irving quite well, first through years of correspondence after he had returned to Australia for good, and finally when he made a final visit to England as a guest of the Vincent Owners Club. On that occasion I was able to invite him to Stanford Hall to see and in fact bid final farewell to the bike. His health did not permit him to ride it and I was unable to ride it either (having hurt my arm in a fall from a lightweight at walking pace!) but Phil Heath who had worked at Stevenage with Phil performed the rites. Afterwards I drove Phil Irving down to Stevenage and we discussed a wide range of subjects involving life and motorcycles in the specially close intimacy that one has on a long car journey. I was left in no doubt that of all his design work the O was nearest his heart because, given the broad parameters, he had a free hand and a clear sheet of paper to work on.

For once he did not have to use up left overs from previous models or use castings and fixtures from previous productions. A rare treat for a motorcycle designer. Because the Roarer was such a technical breakthrough with its geared crankshafts to eliminate vibration at source (and make possible shaft final drive without torque reaction problems) Irving was instructed to follow this layout and this provided the opportunity to design a motorcycle that would at last approach the sophistication riders dreamed of, a motorcar on two wheels. It was the secret dream of the Goodmans too but they knew that in the conservative buyers' market you could not sell a really new design without demonstrating its worth by racing. Cost and complexity did not matter for the Roarer; on its success depended the future of the O and at that time the long term future of the firm. Cost was all important with the O and this is where Irving drew heavily on car practice.

"The average motorcyclist wanted an engine which would develop about 30 bhp, enough for about 90 mph at no more than 6,000 rpm but with plenty of low speed torque for real flexibility. He also wanted easy starting and the best way to get that at the time was to use a car type coil ignition with built in advance in the distributor.

"If the Aspin rotary valve had worked as well as its inventor claimed it might have been used on the Roarer and the Model O but after wasting a lot of time and money on it Willis was disillusioned and went back to ohc, while pushrods were the obvious choice for the O." (He did not make the point, but in fact the single camshaft was so high in the crankcase and the alloy pushrods so short, that it could be described as high-camshaft after the fashion of the M series though in this case it is driven in two stages, the first through the familiar helical gears and the second a light duplex chain tensioned by an adjustable slipper.)

Car makers had cut costs of popular cars by designing for easy production, engines that could be assembled by trained but unskilled labour from simple mass produced parts instead of the labour intensive methods of the motorcycle world. Irving set out along this design path and was encouraged by Eugene Goodman, the production engineer who would have to make. For the first time that I recall he turned away from the previously accepted principle that an owner (probably with the aid of a wedge and a coal hammer) could split flywheels and replace a big end! Car owners did not regard this as essential but had become more interested in mechanical silence, smooth running and long intervals between remedial attention. (Regular servicing was a practice developed by the Services during the war.) Nor were they interested in high power at high revs, but wanted power spread over a wide range to free them from the burden of frequent gear changes.

Accordingly, the Model O engine followed the general principles of the Roarer engine (which employed certain car features with its longitudinal layout and crankshaft mounted clutch) but a specification more akin to grand prix cars (there had been in the vintage period a Continental Grand Prix design with geared crankshafts which may have suggested the design) but leaned more to mass production car practice. The heart of an engine is its crankshaft and here

Powerplant of the Model O. Car-type distributor and type 6 carburettor are visible, as are the welded up front engine plates

Irving cut costs by having his crank webs (shaped like those of the Roarer and like those in then popular Villiers two-strokes) flame cut from thick steel plate (armour plate he had in mind) and surface ground flat. Holes for the pressed-in crank and crankshaft were 68.25 mm apart (same stroke as the MOV) to suit existing precision boring machinery in the works. An extra ½ inch hole was bored through the crank cheeks to take a lining up bar for accurate assembly. Crank pins were 1in diameter over which a hardened sleeve to form the bearing surface for the plain big ends was lightly pressed.

With pressure-fed built-up shafts there is always the problem of ensuring that feed holes from the crank to the crankpin line up after assembly but Irving overcame this problem neatly. The oil passage in the crank emerged on the inside edge of the hole for the crank pin. Here the bearing sleeve was undercut by approximately 1/8in to provide a circumferential channel when the crank web was pressed up. A drilling parallel with the sleeve took oil to a central outlet hole. No need for accurate lining up, no radial holes to weaken the crank pin, and cheap to make. (The same principle was used on the MOV.) Interference fit of main-shafts and crank pin was 3 to 4 thousandths of an inch and for extra security taper plugs were pressed into a boring in the crank pin. Already Percy Goodman had been an enthusiastic exponent of pressed-in shafts to save the cost and loss of strength inevitable with the old taper and nuts. It was no longer a "back yard hammer and wedge job if we had to change a big end" but Irving thought about the service angle. "The con rods were delightfully simple, an RR 56 alloy forging with holes bored at each end to run direct on crank pin and gudgeon pin. Replacing them was no job for the amateur but any engineering shop with a press could push the crank pin out (pushed right through because the bearing sleeve was only a light fit and would support the cranks without special jigs) and although there was enough 'meat' in the rods to permit a bush they were so cheap to make I envisaged new rods and sleeves would be fitted." Following car low-cost design principles all bearings in the engine were plain bushes conferring long life and smoothness (this is immediately apparent when the machine is in use). They have only given trouble through misfortune when run without oil in the sump.

The original design called for a cast-iron cylinder block but the prototype was fitted with a light alloy block, presumably for ease of manufacture. This has pressed-in cast iron liners. Because of the expansion rate of the two alloy components and the use of through bolts from crankcase to cylinder head it is important that the top nuts are only lightly tightened. There are four nuts per bore and the outer ones being ready to hand (the others are inside the rocker box) it is tempting to tighten them to stop the oil weep from the head joint but it's ill advised because this has led to distortion in the past and the consequent weep. Irving, relying on memory, thought a cast iron barrel was fitted but I am sure he was mistaken and when I queried it with Bertie Goodman he said it always had an alloy barrel and gave me a spare alloy one (less liners) to prove the point. I am satisfied that in production there would have been a cast iron barrel and no distortion problem.

I could not resist asking Irving if he thought the Roarer would have lived up to the factory's hopes. "Not without liquid cooling" was his reply. I tried to start a debate over the pros and cons of the design and whether it could have been successfully adapted to atmospheric induction but he changed the subject. It was history and he obviously had no interest in it.

On another occasion I remember he propounded the theory that an ideal design for road racing was a small V-twin. Slim, simple, good balance, plenty of room for separate carburettors (he was thinking of a wide angle twin), easy to air cool and with a narrow and therefore stiff crankshaft. He was at that time intrigued by the possibility of staggering the crankpins to produce different firing angles. He must, I felt, have been pleased to see the success

The Model O at Stanford Hall where it is usually on show

of the Morini V-twins and the Ducati Pantah wide-angle twins. Encouraged, I voiced my own belief that the V-twin was the only engine that really fitted the requirements of a motorcycle... so well that a frame was unnecessary. I thought this reference to the frameless Vincent would persuade him to agree with me but he was never one to look backwards when contemplating the future. "There may be new power units in the future which will present entirely new problems" he said. So I changed the subject.

Car engine practice of containing all the lubricant within the engine, there being no external pipes and tanks, was another prophetic feature. Although it had been used on one or two innovative designs in the past (the original Ariel Four) and the first Triumph twin (the Val Page design) it was not followed on the Roarer which used the dry sump system, albeit with an oil tank forming the front cover of the crankcase. The Model O was a wet sump system with oil pressure fed to the vital points but falling into the sump beneath, to be picked up by the scavenge pump and recirculated. Providing the joint between the necessary detachable sump is above the oil level there is little danger, given a secure drain plug, of oil leaks on the garage floor. It was a practice followed by Oriental designers as soon as they began to produce multis and is largely responsible for their oil retention capability. It was completely satisfactory on the Model O until Eugene Goodman whacked it up to 90 plus on a long downhill slope and ran a couple of main bearings. It was established that under those extreme conditions so much oil was being whirled about by the cranks that not enough was falling to be picked up by the pump. Increasing the capacity of the sump by extending it rearwards part way under the transmission housing cured the problem by providing more reserve of oil.

Although it might be thought that keeping the oil in the crankcase instead of returning it to a separate tank might overheat the oil, and it might do so on a conventional non-unit engine with a small crankcase, the considerable area of the alloy barrel-shaped crankcase, clutch housing and gearbox provided a large cooling area and exceptional 'heat soak' to cope with exceptional conditions, and the unit runs remarkably cool. Provision was made for an external 'throw away' oil filter of car type but not provided on the prototype. There was one point about the lubrication system on which Udall and Irving chose to differ and as Irving constantly referred to it in articles and conversation they may have crossed swords. It concerned the lubrication of the vital gears coupling the crankshafts. Udall aimed a jet of oil at the point at which the teeth meshed: Irving aimed his jet at the point at which they separated. I would myself have chosen Udall's method as most likely to keep the gears well lubricated and quiet but after being sternly advised by Irving that it is likely to cause excessive strains, overheating and damage to the surfaces as the oil is trapped and forces its way out I will no longer contemplate it.

Through lack of hard facts (no information was officially released by Veloce), and sadly no exploded drawing was ever produced, many myths have sprung up about the machine. The most common, probably because it sounds so dramatic, is the story that Eugene Goodman was not satisfied with the steering of the prototype and took drastic action to improve it. The story goes that he cut a section out of the top tube of the frame, had the two ends brought together by straining them with a cable and butt welded. There is evidence that the front fork girders have been lengthened at some time but there is no sign of alteration to the top tube. Bertie Goodman, when quizzed said that was not significant, a new top tube would have been fitted if the crude experiment had proved successful. More recently Peter Goodman said the top tube surgery was carried out on the spring-frame MSS prototype, not on the Model O, which of course he rode on numerous occasions. Irving always maintained that there was nothing wrong with the steering and no alteration was made while he was at Hall Green. Frankly, I don't know the truth of this story although there is evidence that the front engine mounting plates have been cut and re-welded.

Now if the top tube had been shortened it would have pulled in the front down tube and the engine plates would have needed to be shortened... does it matter? Though cost was not considered in the design of the Roarer it loomed largely in Irving's brief and it must have pleased Eugene in particular when he managed to incorporate the internals of the MOV gearbox by an alteration to the sleeve gear to carry the clutch hub and to the layshaft to extend it to take the forward universal joint. Although the clutch followed car practice in being carried on an outside flywheel, Velocette practice was followed in the use of 18 MSS clutch springs and a self-aligning ball thrust race pushed by a hinged plate (also car practice).

Velocette determination to preserve the mystery of their clutch operation was followed though I am sure it was unintentional. The hinged pressure plate was in turn pushed by a rack in a tunnel in the gearbox casting and the pinion engaging with it was turned by a small lever on top of the

No way to treat a lady! A young Peter Goodman tries out the O on a scrambles course

box. An external clutch cable anchored to the rear of the casting pulled on this lever. Apart from the external lever there was nothing to indicate what went on inside the gearbox, no more than on a conventional Velo gearbox. But this external lever was not held down by a nut or stud or conventional fastener but merely held in place by a spring blade, like that for instance used on a magneto contact cover. Curiosity overcame me one day and I prized the blade aside and lifted out the operating arm. It was dogged into a small pinion and as I lifted it out there was a soft 'plop' and something, it was the rack actually, slid out of sight towards the rear of the gearbox leaving the pinion turning idly. A cold chill struck me. I could foresee that as a result of my idle curiosity I would have to remove the engine and gearbox unit, separate the transmission part and in all probability strip the gearbox. I went and made myself a cup of tea.

On return I noticed that on the rear of the gearbox part of the casting there was an anonymous or innocent looking set screw in line with the tunnel in which the rack had disappeared. I removed the stud and with the 3 inch nail which every dyed-in-the-wool Velo man always carried gently pushed. Wonder of wonders, the rack slid into view.

I anchored it quickly by engaging the dog clutch of the arm with the pinion and breathed again. I still ponder if the thoughtful Irving figured some foolish person would make the mistake I had made and had provided the way out. I never had the nerve to ask him. He was not a man to suffer fools easily.

Taking the engine out, which is necessary for any work beyond a head and barrel job, was made as easy as possible. One long bolt running through the front engine mountings lets go the front end. Split clamps around the frame cross-member are just as on the rigid MOV and MAC except that a thin rubber sleeve is interposed to complement the limited rubber mounting provided on the front mounting by rubber washers, like tank mounting washers. Absence of vibration means little is required to keep the unit in place and no movement on the mountings is noticeable in use. With some five nuts removed the engine unit can be lifted out. It takes two to lift it really, and, of course, the cables and wires have to be detached. The cross-over mechanism which allows a normal kick-start crank (normal by modern standards because Irving invented the idea of letting the whole crank pivot at the base instead of just the pedal part which wore wobbly loose in no time) on the offside to operate on the nearside crankshaft detaches itself as you remove the engine, having a spring loaded dog drive.

Although I originally thought that the rubber washers on the front engine mountings were merely stuck to the small engine plates it appears they were bonded to it, being an early example of a new technique developed by Dunlop, and Irving thought the O was the first example of a motorcycle engine with flexible mountings. Certainly I cannot think of an earlier example but it seemed logical, with the design following car practice so closely. Another example of car practice was the use of steel brake shoes of the type which has become almost universal in the car world after the car makers gave up making their own brakes in favour of buying them out from specialist firms like Lockheed and Girling. It was all part of the need to save production costs to keep the price of a sophisticated machine down near to a level that had been established by conventional models. And although the purist might turn up his nose at steel shoes instead of alloy castings they have in service many advantages. They do not corrode, making binding of the linings more reliable, they withstand rapid relining, and the spring pegs do not break off.

The steel brake shoes worked in cast iron drums which were part of an Irving design of front hub first used on Waycott's final ISDT outfit to provide QD interchangeable wheels. The alloy spool centre was trapped between the brake drum on one side and a spoke flange on the other by through bolts. With a brake drum each side it was the design he had thought up when building duo brakes for Vincent HRD and it was revived in miniature form for the early LE machine.

Much mythology has arisen over the coupling gears and I am often quizzed about them. They are in fact very much the same as those on an Ariel four, in size and in tooth form, and it is not surprising to find that they make a similar sound at low speeds, but having had experience of Ariel fours I would say that the Velocette gears are quieter, emitting more of a clicking sound than a clonk. Ariel gears are of course made of steel with, on the later models, a fibre ring riveted to the face to quieten them. Irving chose heat-treated high tensile cast iron (again breaking new ground). Percy Goodman apparently did not like the material although Irving did not know why unless it was difficult to machine, and had another set made in steel which were much too noisy. A set made in Tufnol, a resin bonded fabric material, very popular for car timing wheels, lasted less than 100

No scrambling this time! Fifty years after his last ride on the O, Peter Goodman meets the bike again; moments later he was riding it....

miles before stripping, and a set in soft cast iron were no quieter than the hard ones and caused the first serious 'blow up'.

Frank Mussett, the Australian TT rider, had been dispatched on the O on an urgent mission to Dover (to collect a gearbox needed for Stanley Woods' TT bike) and was no doubt giving it some stick when the steel gears shed some teeth and locked the engine so suddenly that it twisted a crankshaft and cracked the crankcase. When the case was welded up it was discovered that the section was only about half as thick as the drawing called for and the original hardened iron gears were put back and are still there. I detected a certain "told you so" relish as Irving recounted this tale. As the O was a prototype to be tested to its limit a lot of riders apart from the two young Goodmans gave it stick and it's a tribute to the design that it has stood up to so much use and abuse so well.

When I first examined the O there was a spare manifold to take a single carburettor but that went missing before my ownership and I do not think it was ever used. The specially made float chamber that sits between the two type 6 bodies is very neat, but Irving admitted that it might not have been such a good idea for a sidecar machine where on bends surge could have caused starvation of one cylinder. Although a sidecar ratio set of bevel gears was made and found to be a little noisy no one can remember a sidecar being fitted. Which puzzles me because there is an eye bolt for a sidecar seat stay and the roll on centre stand has at some time been removed (and lost), as if its lugs were wanted for a sidecar fitting. It might have been intended to fit a sidecar during the war years when it was a factory hack, and preparations were made but later abandoned. Though the seat height is low by modern standards and I can plant my feet on the ground it is rather a lump to manhandle and one day I might fit a sidecar. It's got plenty of torque to laugh away a modest sidecar and absolutely no "power step" whatsoever. Cam form is believed to be early MSS and Irving said that higher compression and a "smarter" cam could have provided 100 mph performance.

Irving maintained that although the steering at high speed was not quite up to Velocette standard at first, after the forks had been lengthened and a stronger fork spring fitted and the length of the links altered, it was completely satisfactory.

To this I can testify and in fact it was the exceptional stability of the machine which started my love affair. It might at first experience be thought to be heavy steering and with 3.50 x 19in tyres front and rear (interchangeable, remember) it could hardly be nimble but you soon adjust and it inspires such confidence that you ignore all pot holes and road markings and can take a hand or even both off the bars to adjust your goggles without fear of the dreaded tank slapper.

I was still running the machine in, having been assured that the big ends had been re-bushed by a firm of specialists before the engine was rebuilt by a very capable engineer in whom I had complete confidence, when I was invited to ride it to Brussels as part of the 'Ride into Europe' part of the Common Market celebrations. A leisurely ride in convoy seemed a good way of putting in the miles and was most enjoyable until on the way back to the ferry it began to make ominous noises in the engine room. I nursed it for the final miles and it actually locked up as I rode up the ramp. The big ends had seized. Back at the specialist engineering firm it emerged that the man who had re-bushed them had been given the impression that it was only a museum piece and not intended for serious work and for some reason or other allowed the bronze bushes to be a floating fit in the alloy rods. Running two soft metals together is fatal and the bronze bush had chewed up the alloy rods. The boss of the firm took the job over personally and made steel sleeves to press into the rods and if I should ever need big end liners I can get them off the shelf from any Rover dealer!

What is it like to ride? I always say it's just like a Square Four, only a bit smoother, the transmission as well as the engine, because there is a rubber-bushed universal joint in the drive and no harsh chain. It sounds like a Four because the exhaust pipes are linked under the gearbox, and it has a multi cylinder drone rather then the beat of a twin. The gear change is light but does not like to be rushed or it will baulk. The clutch is a trifle heavy but may have been lighter before the cable stop was moved out of sight under the detachable bonnet. This is of course another car type feature: a steel cover hinged down the middle like old time car bonnets, or removable complete, which hides the distributor and direct drive dynamo. Then there is an oil filler on the crankcase and a dip stick. Starting is very easy when the unit is warm although the kickstart ratio is typically Velo-ish in being low geared and there is very little flywheel weight to spin. Twin quick set throttle stops allow cold or hot start positions and with coil ignition to give a good spark, a lazy stroke of the kick start pedal is usually enough. With the throttle stops set back it will idle contentedly like a quality car.

The suspension can best be described as firm. It copes pretty well with serious bumps but doesn't waste its energies dancing up and down all the time. You can of course adjust the friction damping of the girder forks by turning a knob, but there is no adjustment for the friction damping units at the rear. They were preset on assembly and still seem to be doing their job. There is of course no oil to leak. Just as well there is rear suspension, for the seat is a disaster. The saddle part has tension springs in its mattress but no support springs and it's little better than sitting on a board. The pillion portion is no better. Yet Veloce invented the dualseat (although they only saw it as a racing seat) and the prototype monocoque MSS has a shapely two-level sponge rubber cushion dual seat. Strange.

I've never checked the fuel consumption but it's reasonable... about 50 mpg on a run. Would it have sold? Debateable, as I've said. Only in small numbers at first, I think. It would be competing for the luxury market with the Ariel Four, not a big market, and later on it would have been in competition with Sunbeam (not much of a contest) and the BMW.

It would have appealed to the mature enthusiasts who could have afforded to pay a premium price for a sophisticated machine. Younger riders would have gone for something more sporting and never mind a few vibrations... they went for Triumph twins and the clones that followed. But by the late fifties these riders too had matured and were heartily sick of the vibration which shattered cycle parts and limited cruising speeds. Up-dated with tele forks (as a sop to fashion), hydraulic rear dampers (fashion again) and

Titch hooks up into low gear. Just visible is the safety guard for the cable on push-pull twistgrip, described on this page

improved styling it could have been the first Super Bike.

I never found out who invented the unique push pull throttle cable used in the later thirties on the road models. The inner throttle cable was double wound and very similar to a post-war speedo cable inner, the outer throttle cable also being like a speedo outer. Secured to the special throttle slide at the carburettor end, it could be pushed or pulled, no throttle return spring was necessary. The twistgrip was as light as a feather and stayed put in any position without a friction spring. Delightful to use but when the inner became tired it could buckle up where it left the outer at the twist grip and leave you with the throttle stuck open. It first happened to me on a GTP when it did not really matter but ever afterwards I fitted a strip of metal an inch or so long and 3/8in wide above the cable so it could not buckle up. You can see it on the Model O where a stuck open throttle would have been no laughing matter. Push-pull did not appear after the war.

Another Phil Irving design: prototype swinging arm MSS as tested in the early war years by the factory. Note attractive stressed-skin rear end (to be used on the Model O) and adjustable suspension, employed on all later Velocettes, and two-level dual seat, retained briefly after the war

Roarer stands down and Germany takes the Senior TT

Velocettes' supercharged twin was designed and built in eight months

This is the Roarer as refurbished by the works for post war show purposes

When the supercharged BMW and the supercharged Gilera fours began to gain mastery over the outdated but more rideable British singles — Meier won the European 500 cc Championship in 1938 on the BMW and Serafini the title in 1939 on the Gilera — Velocettes did not shrug it off as a fluke. They saw the writing on the wall, saw that the days of the single were numbered, and that a new conception of racing machine was vital if British supremacy was to be upheld. Nortons and Velocettes had for years undertaken the task of upholding British prestige at a cost which must have been distressing to the shareholders. Nortons, preparing for war work, decided to give up the struggle but Velocettes took up the challenge and got out the drawing board.

What they produced was in many ways the most advanced design that has ever been seen in the field. Even more remarkable, it was designed and built in the few months between the winter of 1938 and the TT in June, 1939. They were not so sanguine as to intend that it should be raced in 1939 but they had it roadworthy enough for Stanley Woods to try it out in practice and pronounce the handling ok.

In an nutshell, the Velocette supercharged twin was a vertical twin with separate contra-rotating crankshafts in line with the frame and unit construction of gearbox and final shaft drive. This layout, completely revolutionary, was the result of a lot of logical thought. Other concerns faced with the problem might have drawn more heavily on previous designs... the Gilera layout was by no means new then and, as it happens, is still employed today by successful racing multis... and it would have been comparatively easy to build an across-the-frame twin or four. Not Velocettes. They went much more deeply into the problem.

A four, they decided, was too wide or too long, whichever way you mounted it, and the BMW had proved that the requisite power output could be obtained from a supercharged twin. They admired the shaft drive of the BMW because it was such a clean design and, more important, kept the rear wheel clear and free of oil slinging from a chain. This was one of the problems rearing its head in racing. Tyres were reaching the limit from a traction and life point of view... there was no such thing as cling-rubber then and a first consideration was to keep oil away. The trouble with shaft drive was that it implied an in-line engine; the power loss of another right-angle bevel drive was too much of a handicap. An in-line four engine meant handling problems. It was apparent that the BMW suffered from inherent forces which made it difficult to handle. There was

Close-up of the Roarer engine unit

much learned argument at the time about gyroscopic effects, torque reactions and inertia reactions but Velocettes seemed to have decided that the gyroscopic forces were the only ones to worry about.

Inertia forces only arise when the engine is accelerated or decelerated sharply, as when running in neutral, and are not I think important; torque reaction is resisted by the machine itself, tending to twist the frame. Gyroscopic forces are set up when spinning flywheels are disturbed in a fore and aft plane and tend to put the whole machine into a slow roll. The effect is easy to demonstrate and can only be alleviated by reducing the flywheel effect to a minimum and keeping the machine on an even keel. Velocettes decided to eliminate the effect at source by employing contra-rotating shafts which would automatically cancel out each others' gyroscopic tendencies. It meant extra weight but they could — and did — use ultra-light alloys (elektron was the lightest then) and add lightness in all details.

With the decision to have two crankshafts in line with the frame, much of the rest of the design followed easily. One crankshaft could take the drive out to the gearbox and on to the final-drive shaft in an easy line, the other crankshaft could driver the supercharger. In the design of the top half it was natural to draw on KTT experience, and although the cylinders were cast in one the separate cylinder heads closely resembled KTT components. The only real difference, apart from the shaving of the fins to enable them to fit snugly, was that the rocker boxes were bolted on instead of cast in one with the heads. Overlapping hairpin springs, as on later Vipers and Venoms, were employed. The individual camshafts for each cylinder were driven from a common top bevel; the drive shaft, which was waisted to give it some springiness and relieve the load on the Oldham couplings, ran between the two cylinders, and the lower driving bevel was on a short shaft running backwards from the timing case in front of the unit. This was driven by spur gears connecting with the off-side crankshaft. The oil pump, which fed vital parts through jets as on late KTTs, was skew-driven from the near-side crankshaft. Although it was designed before the days of streamlining, the importance of reducing frontal area was realized and the front of the unit was faired off with a bulbous cast-alloy oil tank which formed the timing cover.

Crankshafts with bob weights — like those on early Villiers engines — were employed instead of a full flywheel as the total rotating masses, including the coupling gears, gave adequate flywheel weight. Shaft and bob weights were forged in one piece and the crankpins, which took standard KTT big ends, were pressed in. H-section steel rods carried clipped on oil pipes which fed oil to the little end whence it sprayed out on the underside of the piston crowns. This was a scheme used on some of the works singles (the 500 cc models) to cool the piston, but was later abandoned.

Pistons resembled KTT components but had an extra compression ring and all the compression rings were wedged shaped and worked in tapered grooves. This was a system developed to prevent ring-flutter and blow-by under extreme conditions.

The two crankshafts were arranged to move inwards at the bottom of their travel so that oil would be flung to the centre line of the sump, and the coupling gears of steel, with coarse-pitch teeth, were fed at the point of mesh with an oil jet. Aft of the coupling gears on the nearside came a multi-plate dry clutch and the gearbox which provided four ratios. The output shaft was above the input shaft. To bring the final drive level with the wheel spindle and keep the shaft straight, the gears were of unusually large diameter but quite narrow. Needle rollers were used instead of bushes for the free gears. The layout and detail design of the final shaft drive and rear bevel are easy to visualise for they were substantially the same as on the LE Velocette... that part of the exercise was certainly not wasted. At the forward end the universal joint was housed in a bulge in the swinging-arm pivot cross tube. On an LE this is a simple Hookes joint, as used on car prop shafts, but on the Roarer it was a very hush-hush constant-velocity joint.

The story goes that at the outbreak of the war the Ministry of Supply desperately wanted a constant-velocity joint as a sample and they borrowed the one from the Roarer. Whether or not it was ever returned or replaced I do not know.

On the other side of the engine the crankshaft drove the Centric eccentric vane blower direct through a toothed

---

Roarer reborn

NEVER in my experience of motorcycle restoration, renovation, rebuilding, call it what you will, has there been an undertaking as daunting as that which faced Ivan Rhodes when he acquired the Roarer in 1983. It looked quite presentable, having been refurbished at Hall Green in 1956 for exhibition purposes, but there was nothing much inside the barrel shaped crankcase, which had housed geared contra-rotating crankshafts, because on that occasion it had been stripped to allow an artist from "The Motor Cycle" to produce an exploded drawing. The heavy bits had been left out when it was reassembled as a show piece for no one wants to lug a heavy bike round the show circuit. It was heavy at 370 lb dry despite extravagant use of electron — super light alloy — and masses of holes to add lightness. John Griffith, journalist, PR man, former Vintage Club President and joint founder with me of the Stanford Hall Motor Cycle Museum, bought the gutted Roarer and several other Velocette historicals or desirables in the nick of time before the liquidator and consequently the scrap dealer moved in. Sadly, John did not try hard enough to find the box in which the internals had been stored and the worried staff at Hall Green could not have been expected to worry about such a problem. John had always hoped that it would be possible to cobble up some internals which would make the Roarer run if not roar and we often discussed possibilities. Leicester engineer and motorcycle enthusiast John Cragg was commissioned to make a start but John Griffith was tragically killed in a motorway crash and the project came to a standstill.

No compromise entered Ivan's mind when he bought the Roarer (for an undisclosed sum, as they say). It had to be rebuilt exactly as it had been built originally, down to the last nut and with metal to the original specification or better. The trouble was there were no drawings, merely the exploded artist's drawing. Technology being what it is now, the saying "what man has made man can

A Saga in itself! Ivan Rhodes and son Grahame who had to recreate most of the complex internals of the Roarer with nothing to help them other than an artist's exploded drawing

coupling and an axial shock absorber. Velocettes experimented with a blower of their own construction but gave it up. The development programme began with a low boost of 4 lb with a comparatively high compression ratio of 8.75 to 1 and ended with high boost (13 lb) and low cr (7.5 to 1). The carburettor, a 1 1/16 in TT Amal — very small by present-day standards — was mounted above the blower on a right-angled cast elbow. The magneto was tucked in behind the cylinders and driven by a skew gear mating with teeth cut on the circumference of the blower drive coupling. The most unusual feature of the layout was that the cylinder heads were reversed so that the exhaust ports faced to the rear and the short, slightly megaphoned, pipes ran straight back at high level. The inlet pipe ran upward from the blower, over the cylinder heads and then branched to feed the inlet ports. This gave a nice long inlet pipe to damp down

---

make again" is common but that presupposes you have a pattern or a drawing to work to. Ivan had, however, certain advantages. His eldest son Grahame is a talented engineer draughtsman, his younger son Adrian is an engineer.

His specialised metal finishing business had brought him a host of engineering contacts, not the least being Rolls Royce. He also had an understanding wife, understanding of the nocturnal habits of motorcycle engineers when immersed in a project. The Roarer was unbelievably complex and unlike anything else. It must have cost a great deal to build and one cannot but admire the Goodmans' courage in putting all their racing money on 'one horse', for to build say two more and spares to make a reasonable team would surely have been beyond their finances. No mention seems to have been made of making more.

It took a year of Grahame's spare time to visualize, measure, calculate and draw the missing parts. It would have been much easier to have designed a similar engine from scratch. Then patterns had to be made and parts forged and finally machined mostly by Grahame in the Rhode's workshop. One part, the skew gear drive for the magneto taken off the muff joint between a crankshaft and the supercharger, took a year to complete after Grahame had made the blank. Only the Rolls Royce engine factory at Bristol had a machine tool which could cut the special tooth form and Ivan had to wait a year because it was in constant use. When the Roarer (incidentally, so nicknamed by Willis because of the flat, raucous note of the short exhaust pipes) eventually fired up first time, the six year slog had been worth while! Its official debut was the Historic Parade in the Isle of Man in 1989 with Stanley Woods there to watch. Unfortunately, a big end tightened up and Grahame pulled in at Ballacraine rather than do more damage. At the time of writing the engine was dismantled to attend to clearances and lubrication. After all, with testing and running in miles at Cadwell and demo laps of Mallory Park the Rhode's Roarer has probably covered more miles than ever it did for Velocettes!

Left: 'Start up Day', May 7th, 1989. Left to right: Len Udall, Grahame Rhodes, Ivan and Renee Rhodes, Peter Goodman. Stanley Woods is on the Roarer

pulsations and a very efficient straight out exhaust but was by no means ideal from a cooling point of view. The truth of the matter was that Udall intended to use liquid cooling with Glycol but when the work was well advanced... too advanced for a complete turn round of the engine... Percy Goodman said no. Liquid cooling, he said, was too way out for production machines and the Roarer was intended to pave the way for a road going twin-crankshaft twin.

It is said that the sight of the liquid-cooled supercharged four-cylinder AJS with its tangle of pipes frightened him.

Cooling did provide problems in the development stage. There was trouble with valve sticking and the stems were chromium plated for this reason. When the machine appeared for trials in the Island air scoops were fitted to deflect air on to the exhaust hot spots.

The complete machine was typically Velocette in its black and gold beauty, not unlike a KTT in general appearance. It being 1939, the forks were the specially developed Velocette girders with a guide for the fork spring and the rear suspension was pivoted fork with air-sprung and oil-damped units. (it always seems unfair to give credit for present-day pivoted-fork suspension to the McCandless brothers, as many people do, when it is clear that Velocettes originated the system.)

The frame was duplex with widely-splayed down tubes to straddle the crankcase, and the bottom rails to which the engine was clamped at crankshaft height were detachable to permit the complete power unit to be removed quickly. Brakes were full-width in elektron hubs, and Venom brakes would appear to be directly descended from them. Wheel sizes were contemporary, 21in at the front and 20in at the rear. I have been told by someone who was close to Velocettes at the time that the question of rear-tyre wear reared its ugly head. Although in the Isle of Man try out the engine was only producing 38 bhp — approximately the same as the factory singles — it was feared that it might not be able to get through a TT without a rear-tyre change which, of course, would have been out of the question. It is quite likely that the smooth torque and vivid acceleration of the blown multi would promote enough wheelspin to scrub out a normal tyre.

One curious fact — and an encouraging one — did emerge in the testing. The Roarer, though it only produced the same output on the brake as the single, was actually 6 mph faster on the road. This was attributed at the time to the improved streamlining of the engine unit. A nice theory but not one to which I readily subscribe for although the front of the crankcase was faired off by the bulbous oil tank the frontal area of the complete unit was considerably greater than that of the single which, of course, had the traditional Velocette "slimline" crankcase. I am more inclined to think that the superior torque and wide power band (2,000 - 7,000 rpm) as against 4,500 - 6,700 of the single, plus the greater acceleration, enabled the twin to get up to its maximum power quicker in the available test area. In other words, the single was probably not reaching its optimum maximum.

Anyway, a maximum speed of 130 mph was pretty good for a first outing of a complex and in many ways revolutionary power unit, and the calculated target maximum of 145 mph was not too far away. The blown Gilera Rondine was reported to produce 50 bhp and reach 140 mph and the BMW not far short of it. With the twin back on the test bench, where the blower could be driven from an independent source, to experiment with speeds and pressures, Charles Udall upped the output to 54 bhp in two months... the two months before the war put a stop to the enterprise.

Inevitably one feels like speculating on the possibilities of the engine in unblown form. Post-war research into induction and exhaust characteristics raised the speed of unblown engines to about the same level as pre-war supercharged ones and it seems likely that had Velocettes had the resources to redesign the Roarer for atmospheric induction... with the exhaust ports forward... it could have been competitive for a number of years.

But as it happened, they were concentrating on the LE and striving for export production of the push-rod singles and there was no time or money for serious racing.

The Roarer did not alas, get to the starting grid in 1939, and, because superchargers were banned by the FIM, when racing was resumed after the war, it never ran in anger. It was designed for forced induction with the blower as an integral part of the engine unit and could not easily have been converted to atmospheric induction. The war over, there was another war to win, the struggle for economic survival, and the race for exports; the prizes were dollars, not racing laurels. Velocettes, like our other manufacturers, had to forget racing for a while and concentrate on bread and butter machines and things would never be quite the same, for Harold Willis, who had devoted most of his working life to developing racing Velocettes, was dead, stricken down by meningitis at the very time the Roarer was being tried out in the Island. It was a late arrival, too late to be regarded as anything more than an experiment. At least Stanley Woods was able to report that the handling was first class — the best he had experienced.

With the Roarer a non-starter, Velocettes relied on the faithful 500 singles unchanged from the previous year. Nortons, officially retired from racing, let Daniell and Frith borrow their 1938 mounts but Joe Craig was no longer there to direct operations, for having become redundant he had gone to BSA who, it was said, were thinking of taking up racing with an ohc vertical twin.

The only British machine technically capable of challenging the supercharged BMWs, now ripe for victory after several years' development, was the water-cooled vee-four supercharged AJS. The Matchless concern had struggled doggedly and at considerable expense to get the bugs out of this design, originally air-cooled, but though they had now achieved speed and stamina it was far from being a handleable proposition. So the Senior ran according to form with Georg Meier and Jock West using their abundance of power to get out in front and then take it comparatively easy while Freddie Frith, on his 1938 Norton, rode to the limit to come home third ahead of Stanley Woods on his basically 1937 Velocette. There was a great show of black and gold down the field, much of it cheeky 350, privately owned, Mk. 8 models. Les Archer was sixth on a 500, Mellors seventh. Whitworth again shone as an ascending star by finishing 12th on the 350 he rode (minus a footrest) into fifth place in the Junior.

The Junior was something like a race, and very like the 1938 race, with Stanley Woods putting it across the Nortons again but this time harassed by a screaming five-piston (four working and one compressing) DKW which in the hands of a Woods, Frith or Daniell might well have won.

Stanley was very nearly tricked into losing the race by an optimistic rev-counter.

He wrote his own story of the race immediately afterwards and it provides an interesting insight into the thoughts that flit through the rider's mind as he hurtles round the Island course. Practice had put him top of the Junior leaderboard early on and his final pre-race lap had been little more than a formality except that he was doubtful of the accuracy of the rev-counter. Accordingly, a new one had been fitted before the machine was handed in.

From the moment of flag fall in the race, the Velo seemed to be going well. Down Bray Hill and away from Quarter Bridge towards Braddan Bridge and Union Mills it seemed to be buzzing particularly well. Mentally he put this down at the time to what he called the "special preparation he had taken to ensure that the oil was thoroughly heated during the warming up period." (He was not going to tell the world that they filled the oil tank with hot oil.) According to the rev-counter, he was motoring very rapidly because he was getting peak rpm on a higher gear than he had used in 1938. When he reached his first private signal station at Sulby he got a shock for the signals indicated that he was merely equalling last year's time. Being number one in starting order, that was all the information he could be given at that point but at the end of the lap his Glencrutchery Road lookout information indicated that he was second to Frith and 10 seconds adrift. The only thing to do was to buzz the engine some more.

Peak revs were supposed to be 7,000. He had the old rev counter changed because it was loath to go over 6,800 although the motor seemed willing. Now he found that the clock would go to 7,500 and there was still no valve bounce but he still did not suspect the instrument because they had fitted stronger valve springs before the race. Sulby said he was still 10 seconds behind Frith, but Fleischmann on the DKW was only two seconds behind him Then it began to rain and from Ginger Hall to Ramsey it was so slippery that he played safe and left it in third. Seems other people did the same for he did not lose any ground. But the Sulby signalling station had now been able to do its sums and reported that, at the end of the first complete lap, from their point of view, Stanley had been no higher than fourth behind Fleischmann and Daniell and 23 seconds adrift.

It was at this point that he began to suspect the rev-counter, and leaving Union Mills he put it to the test by letting the revs build up to valve bounce. Seven thousand, seven five, eight thousand flicked the needle before the valves bounced. After that he rode with grim desperation to win back the seconds he had lost. At Sulby he had pulled back five seconds. By the time he reached the pits to fill up he had won back another two seconds and was third. Soon after, the rev-counter packed up completely and after that he took it up till the valves bounced, flattened himself along the straights and tried everything he knew.

Near Sulby he had a close shave. Someone had fallen off and the road was nearly blocked. Stanley, arriving flat in third at over 100, glimpsed a woman waving her hat violently at him. With the second sense which operates when one is keyed up to that point of concentration, he recognised the wave as a warning, not a welcome, eased up and squeezed by.

Sulby had good news: 11 seconds behind Frith, six ahead of the 'Deek'. On the fifth lap the news was even

Veloce photograph, circa 1939, showing the Roarer's crank coupling gears. What is the trials Velo under the workbench; could it be the Waycott ISDT machine?

better, three seconds behind Frith, and by the end of the lap the Grandstand reported first plus two seconds. By Sulby he rejoiced in the news that he was leading by 27 seconds and on the last lap the figure was 41 seconds. Confident, he eased off, very nearly eased off too much, for he had forgotten that the Sulby signals were a lap behind.

He had plenty of time to ponder on this mistake as he sweated it out for 10 minutes in the finish enclosure waiting for Daniell to finish... Frith had retired, and he had been caught out before by Daniell coming up from behind with a wet sail. But he did win by a mere 8 seconds. And I suppose that all the experts figured he had been playing canny all the time. Starting off slowly to fox the opposition and then turning it on as and when suggested by his "secret" signal station.

The 1939 Isle of Man TT marked the end of an era. Our one-time supremacy in all classes had been challenged by more advanced designs from Germany and Italy and finally toppled in the Senior and Lightweight classes. Even in our traditional stronghold, the Junior, there was a German two-stroke knocking on the door and needing but a little more development and an Isle of Man star in the saddle to win. Our only comfort was that it had taken superchargers to beat, us and we still thought that they were a mild form of cheating. One interesting sidelight that year was provided by the ACU weighing all the machines, and some interesting facts emerged.

The lightest Senior mount turned out to be Meier's BMW at 302 lb (by contrast, a standard roadster BMW ridden by Tim Reid weighed 351 lb), yet Frith's Norton came out at 336 lb, which was slightly heavier than Daniell's and heavier than the non-work's machines. The 500 Velocettes ridden by Stanley Woods weighed 338 lb and for no apparent reason the similar mount of Ted Mellors weighed 9 lb more. The water-cooled supercharged AJS looked heavy, and was... 405 lb The Junior Velocettes ranged between 306 lb (same as Daniell's Norton) and 329 lb (this was Stanley Woods' winning machine. Why so heavy, I wonder. The majority of the non-works machines were around 321 lb. The DKWs were around 337 lb and the heavyweight of the class was the parallel-twin supercharged ohc NSU at 369 lb. Of the Lightweights, the Pike Rudges were on their own at around 297 lb, Mellors' winning Benelli weighed 291 lb and the DKW 316 lb.

Some works machines had an extra scavenge pump, a very slim gear pump, in the lower bevel box to evacuate any oil build up in that region. Certainly in the later stages Velocettes were as concerned about getting the oil out of their engines as in the early days they had been to get it in.

Another form of incidental information concerns fork springing experiments in the 30s. Ernie Thomas told me of tests he carried out with a rubber spring fork named the Macbeth. Very soft and comfortable it was, he said, at low speeds but at high speeds it bounced up and down like a yo-yo. I've turned up a description of the Macbeth fork (named after its inventor, Colin Macbeth, a pioneer of rubber suspension), and find the rubber was used in torsion, being bonded between two steel discs mounted in the fashion of girder fork friction dampers, one disc anchored to the top crown, the other to the fork link. No fork spring was used.

It was claimed that the hysteresis loss... the energy lost in distorting the rubber... made friction or other forms of damping unnecessary. This is where, in my opinion, the designer was wrong. Rubber in torsion does have an inherent damping effect but not enough to cope with high-speed operation. I think riders of early rubber sprung Greeves models would agree. And I recall carrying out tests with a rubber-in-torsion suspended machine which had no auxiliary dampers. It gave a wonderful ride but when you hit a really big bump it swallowed the bump and then a moment later it catapulted you in the air.

Another suspension tale Thomas told me concerned the Brampton girder fork. Velocettes had always used Webbs on the four-stroke models (for some reason best known to themselves, the very successful Mitchell brothers used Bramptons) but Bramptons did not give up trying to get the contract and in the 30s their representative persuaded Velocettes to let him fit a pair of his forks for test. Unfortunately the test forks were reamed to too-close limits and the spindles seized. All efforts by the embarrassed representative to free them failed and finally he had to saw through the links before the forks could be removed.

Looking back through the racing history of the KTT Velocettes, one can detect something of a hiatus around 1936 when no new KTT machines were made... apart from the solitary Mk. 6; and the work's entries were, apart from the short-lived dohc, the three-year-old Dog Kennel jobs. This state of affairs was in part due to an abortive excursion into the realm of rotary valves. It was really more of a rotary conical combustion chamber than a simple rotary valve and certainly looked promising. One could visualize the charge being whirled round, fired and then flung out of the exhaust port. The designer claimed astronomically high revolutions on frightening compression ratios while using low-grade fuel and cooking plugs, claims which characterised other rotary valve designs from time to time. Velocette obtained a development option on the Aspin engine and spent a great deal of time and money before ending up with an engine which was not very reliable and produced no more power than their conventional jobs.

All thoughts of getting the Roarer really "sorted" for the 1940 TT soon had to be forgotten. War work became top priority and Velocettes turned their skills to producing components for the Air Ministry. It was a task for which Eugene Goodman, the production engineer, was particularly well equipped. The next years were long, hard ones and not always rewarding financially. Mr. Eugene recalled with some bitterness an occasion when they were subcontracting the manufacture of a certain aircraft component. He evolved a system of manufacture which cut the cost considerably and dropped his price accordingly while still allowing a working profit. It was not long before a man from the Ministry turned up and cut the price again below the economic level. The firm would have been better off if they had carried on with the original method.

Motorcycle manufacture continued at a lower level through the production of a militarized version of the MAC model. The MAC as it stood was quite a suitable service machine. As a matter of fact the first military MAC models were intended for the French Government. But the batch was taken over by the Ministry of Supply after the fall of France - but Velocettes soon added a number of refinements which made it a very coveted machine among service personnel (most of them went to the RAF). The engine and gearbox were unaltered save for the use of cast-iron timing covers and gearbox end plates and the ATD was retained. Scorning the steel plate crankcase shield of other WD models, Velocettes built a massive cradle lug into the frame which formed a built-in crankcase shield. A lot of thought went into other details. Racing-type rubber bump stops were fitted to the front forks, and a new and very simple foot change pedal linkage was devised. The rearward extension

Velocette in khaki — a military version of the pre-war three-fifty MAC

of the pedal had a right-angled sleeve which engaged directly with the operating arm on the gearbox, by way of a ball and socket.

The usual clevis links were eliminated by this simple universal joint arrangement and the gear change was clean and crisp. The foot brake pedal was remounted independently of the footrest on its own frame lug and well tucked in. Another detail difference in the MAF model was a new design of kick-start crank in which the crank had a fixed pedal instead of a folding one but pivoted in a boss on the kick-start spindle. The effect was that of a folding pedal but the bearing surface was greater and less likely to rattle.

The normal bolted up rear stand and mudguard extension gave way to spring clip fittings in the interests of quick puncture repair, and a nice little rider's touch was the extra spring clip to hold the hinged mudguard extension aloft while removing the wheel.

Instructions for adjusting the clutch were clearly written on the final-drive chain case — an elaborate case enclosing the lower run of the chain. I supposed it was just as well that the MAF model, was not normally issued to the "brutal and licentious soldiery". It was difficult enough instructing some of them in the adjustment of a conventional clutch!

Altogether some 5,000 MAF models were made, a trifling number compared with that of other WD machines, and they are now extremely rare.

Arthur Lavington

THE NAME E.A. Lavington (*writes Bill Snelling*) was almost invariably matched with Velocette in events' programmes, be it at the TT, where he flew the Veloce flag with his Mk VIII KTT, or as a front runner in South Eastern Centre ACU road trials.

Arthur's ... the E. was Edgar ... initiation into all things Velocette occurred at the Shepherds Bush, London firm of L. Stevens, of Goldhawk Road. Stevens was something of a passing-out college for Velo dealers. Along with Arthur, Geoff Dodkin was also employed there. Geoff recalls that the weeks preceding the TT were very unproductive in the Stevens' workshops, for all hands would be enlisted in getting Arthur's bike prepared and assembled in time for practice week. Arthur's first Island race was the 1949 Junior Clubman, on a KSS; the bike ended in a heap at Hillberry when the girder fork spring broke as he approached the sweeping right-hander.

In the early 60s Arthur started his own workshop, first in Earlsfield and later in Tooting, South London, in a small alley between a baker and a radio shop. The 'workshop' was the ground floor of a very small cottage, and the garages were the old stabling block for the King's Head pub across the road. The upper floor was the spares department, a veritable treasure trove of bits for Velos of all ages. It held the biggest stock of KTT parts around, but one had to be well-in and well-vetted before one could persuade Arthur to part with any of those bits!.

Arthur's garage was second home to the many mobile police force member's stationed at Tooting nick; you could almost hear the massed-whisper of the LE brigade heading our way when the kettle went on! The cops had their own workshops, but Arthur's place was handily situated if something went amiss that couldn't be explained away by normal wear and tear. Like the time two had been playing tag round the back streets, the result being a shunting session, with one bike's legshield bearing the distinct marks of the other's bent pannier box. They found their way to Upper Tooting Road, where we were soon busy replacing the incriminating evidence with new parts... very little on the police bikes were nonstandard. Then another Noddy was heard coming up the slope. This time it was the local sergeant, but he didn't stop for a tea break this time. Slowly circling and surveying the carnage, he muttered 'I see nothing' and drove back to his nick for a brew. Nothing more was heard of the incident.

Even with his own business, E.A.L. was a past-master at last-minute preparation. I recall working through the night to get his bike ready for an Aberdare meeting, then loading it and about half a dozen bods aboard his old Ford Thames 10/12 cwt., and setting off very early in the morning for Wales.

The South Eastern ACU Centre was well known for its road trials, and Arthur was a real-past master in these events, both as organiser and competitor. The 'B. J. Goodman' he organised, run by the Dorking Centre of the Velo Club, was guaranteed to lose a high percentage of the entry, who would be sighted at points over Kent, Surrey and Sussex. The combination of reverse route, herring-bone and clock face made for a lively lunch stop as those who had got it right tried to placate the lost herd who were always certain that the route

Arthur Lavington at Quarter Bridge, 1954 Junior TT

card they were given was wrongly printed

Arthur was nearly always ably navigated by his wife, Brenda, and was always on his faithful LE. It was usually a battle royal for honours among the Lavingtons, the late Bryan Amos, and June, on a Royal Enfield Bullet, and John Gatland, who rode a BSA twin. Occasionally Brenda was unable to compete, and yours truly would then be drafted in as navigator. One such event was a Saturday night and Sunday morning all night affair run by the Guildford Club, starting just outside Guildford and heading west to finish in Wales. We started well, keeping up the schedule until we got in and around the Welsh hills. Then, the combination of 20+ stone, an 8 hp 200cc side-valve power unit and inept navigation overcame Arthur's ability and we gradually dropped behind the 24 mph average required. It was not for want of trying. Arthur was making a real effort when the inevitable happened on a long, downhill left-hander. The Noddy was already cranked on to a footboard when the corner tightened up. We leapt over the ditch bounding the road and careered along the hedge. I was plucked off the bike by the hedge and landed in the ditch, to watch Arthur fighting to halt the wayward bike. This he did, but when he stopped, there was nothing to put his left foot on, and he slowly toppled into the ditch. In a flash he was up, and together we lifted the bike across the ditch and were away again.

A thrashed side-valver drinks juice like a parched Manxman, so it was not surprising that such antics should make us run low, then finally out of fuel, well out in the country, and by this time quite a way behind the rest of the field. There was no one to come to our assistance. In the dawn

Arthur Lavington in the 1960s

light we saw a lorry parked outside a farmhouse. We went to investigate its fuel tank content, siphon tube and can in hand. A quick sniff, yes, it was petrol; we then began liberating a couple of quarts. Arthur was to my left when I became aware that somebody, or something, was on my right. It was the farmyard collie, sitting by my side, silently watching all that was going on. We filled up and were away in a flash, but the dumb-dawg followed us for about a couple of miles, still without making any sound.

Arthur's LE was very standard in the engine department. It started life as a Mk II hand-change three-speed model, but was gradually brought up to Mk III foot-change spec. To cater for Arthur's enthusiastic cornering ability, a pair of Venom fork springs raised the front end, and longer Girling units replaced the standard spring-only units; the rear subframe was double skinned to stop these units punching their way through the bodywork. Otherwise it was quite standard — and its performance a testimony to the little Velo's basic qualities and Arthur's riding skill.

I was enlisted for the Brighton semi-sporting road trial, held up and down the South Down's byways, around Chanctonbury Ring and the Ditchling area. The ups weren't too bad; I think I had to leg it up only one hill that defeated 200 cc of gently steaming machine. It was coming down the other side of the rises that nearly caused a prang. In his enthusiasm to get by a sidecar outfit that was holding us up, Arthur switched ruts and dropped into a rain gully that was deeper than our ground clearance. After tobogganing down on the footboards for what seemed miles, the wheels touched bottom and somehow — I did not see quite how, as I was sitting with eyes and buttocks clenched tight — we cleared the gully and carried on. Arthur's riding was a little circumspect after this incident, but only for the next few miles, then it resumed the usual competitive edge.

Occasionally this edge would cause friction between Arthur and Brenda on road trials. It was clear when something had gone amiss enroute. The Noddy would appear at the finish with the edges of the footboard smoking, the radiator steaming, and you knew to give the prickly pair a wide berth for a time, to let *them* cool down.

Although not a founder member of the Velocette Owners Club, Arthur gave valuable service to the club. In its early days the club existed round the Dorking Centre (at least, that's what the other, mainly northern, centres seemed to think), and Arthur, Brenda and at least half a dozen other Dorking members would be pressed to attend the inauguration of many new centres.

E.A.L. was best known nationally for his riding in the Island on his Mk VIII KTT. In later years the TT became Arthur's sole race meeting, apart from an odd event at Crystal Palace. The Mk VIII came to the Island at least one year, in boxes, to be assembled in the 'paddock hotel'. Many's the time newcomers to the TT course would pass the old black Velo down the straights — it was good for about 108 mph — only to be re-passed time and again through the twiddly bits. After his initial ride in the Clubman's TT, Arthur rode in the Junior TT most years from 1954 to 1969. From 11 rides on the Mk VIII he had five finishes, his best placing being 39th in 1963. His first TT ride netted him a 57th in the Junior, but the Senior ended for him at Quarter Bridge on the first lap. He was convinced that the magneto had failed, but all that had happened was that the main jet had unscrewed itself. Another time he came in after a practice lap feeling unwell, suffering from double vision. The reason was clear when he took his goggles off and the lens from his specs fell out! Another year the clutch was not set right, slipping right from flag-fall. He nursed it round to the Greeba area, but it was useless to carry on like this. Persuading a nail out of a fence to use as the Velo clutch adjustment tool, he carried on to a finish. Arthur was loaned a Thruxton for the inaugural Production TT, but retired after a slow first lap.

The 1969 TT meeting was to be Arthur's last race. He had qualified well enough, and was half way through a round of golf in Noble's Park with fellow Velo men Ken and Doug Law, when he decided to do a lap and stretch in the new rear chain. He got ready and went out for the Friday evening practice. Going through the swervery beyond Kirk Michael, Arthur was clipped by a faster competitor at Alpine Cottage, a mile before Ballaugh Bridge, and hit the roadside wall, sustaining severe head injuries from which he never recovered.

Arthur is remembered by the Velocette Owners Club with their E.A. Lavington Memorial trophy for a Velo's-only race at their race meeting, held first at Llandow and now at Cadwell Park.

E.A.L. at the Waterworks

A motorcycle for Everyman

The LE is a design which for ingenuity, compactness and fitness for a specialised purpose will eventually be seen in two-wheeler history as a masterpiece

LE Velocette: "Never has a motorcycle been designed and built with such idealism"

It was during the long, hard years of war that Eugene Goodman's dream of a quantity-produced "everyman's" lightweight began to take shape in his mind.

Everyone with experience of the conditions after the first world war realized that there would again be a tremendous demand for personal transport. Eugene, a production man first and foremost, was convinced that the traditional motorcycle with its tube and lug frame and multitude of bolted-on bits and pieces could never be produced in the quantities necessary for really economical production. Already he had explored some of the possibilities with the pre-war stressed skin rear frame prototypes and was satisfied that steel pressings could cut cost and production corners.

The requirements for an "everyman machine" were gradually laid out. It had to be inexpensive though not "cheap"... and rather nasty in consequence. It had to inexpensive by reason of mass production, like the Ford Popular car. It had to be light to handle — not necessarily ultra lightweight in weight — so that it could be managed by young and old of both sexes. Easy starting, preferably with an alternative to the rather barbarous kickstarter, was a big must. It had to be economical to run, need no attention between specified service schedules, and be clean to ride, so clean that special clothing would be unnecessary. Above all it had to be quiet, really quiet. Near silent. Noise, it was realized, was the biggest objection of non-enthusiasts to motorcycles. Only the silence and good manners of a car would, it was felt, woo the man in the street and allay his fears that two-wheelers were "nasty, noisy, dangerous things".

Now the interesting thing about this ideal specification is that it was not new or revolutionary even in 1940. Many thoughtful designers and enthusiasts had arrived at the same conclusion in the past, but no one ever carried it out to the letter. Some attempts had been spoiled for a ha'p'orth of tar... the use of a nasty little two-stroke engine (how nasty two-strokes could be in the 20s and 30s!) in flimsy, sounding-board pressed steel frames. Some attempts achieved limited success at first but finally failed because they were compromises. The example I have in mind was the Silver Arrow Matchless which had most of the requirements — the quietness, the easy starting, the good manners and the spring-frame comfort — but for reasons perhaps of playing safe and being certain of a small market, at any rate, retained a normal motorcycle image. And retained thereby most of the motorcycle snags of frequent maintenance and filth. Lacking the requisite motorcycle appeal of performance and noise, it fell between two stools.

The requirements had in fact been set out lucidly in 1929 by that great seer and visionary, Ixion, in *The Motor Cycle* at a time when the Manufacturers' Association was planning to encourage and publicize utility machines by staging an everyman's reliability trial. A trial was held and it served to encourage a rush of pathetically conventional lightweights decked out with legshields and elementary enclosure.

Ixion set out his these under 11 headings, adding typically Ixion asides by way of qualification. It ran like this:

1. Easy starting. (Beyond putting this first on his list he made no comment. It was obviously to be accepted without argument.)
2. A roll-on prop stand. (He did not think comfort could be achieved with ultra light weight so an easily operable stand was made.)
3. Puncture proof. (He made no suggestions as to how this was to be achieved and, although tyres had improved, we are still no nearer a solution.)
4. Silence. (No more sound than a Vere de Vere aristocrat — period touch, this — eating soup. He admitted that he liked a bit of noise himself but the Great British Public didn't!)
5. The minimum of controls and instruments. No ignition or air control. But an ammeter was essential.
6. Good handling. (So good that an old man could ride it on grease without falling off.)
7. Weather protection. (So that one could ride in a clean suit.)
8. Reliability. (To be assumed, though he thought the existing standard was quite high enough.)
9. Economy. (He did not think running costs mattered within reason if the machine was easy and pleasant to use. It did not matter a split pin whether it did 70 mpg or 120 mpg)
10. Electric lighting and horn essential.
11. Capable of being hosed down clean in a few minutes. (This facility seems to have loomed large in his mind for he elaborated thus: "Mr. Ga Ga will finish each run in one of the following uncompromising frames of minds, very thirsty (fine August weather), very pleased with himself after long ride, very frightened after ride on greasy roads, very wet and cold (11 months of the year). In none of these moods will his soul crave for a subsequent two hours in the garage armed with a cask of rags and a dish of paraffin, several brushes and a tin of metal polish. A fire and a bar represented his normal aspirations on getting home. Satisfy these aspirations instead of condemning him to a task which will remind him of the stories he has heard about Devil's Island.")

Remarkable how the 1929 list tallies with the Velocette list of the 40s — yet not really so remarkable, for human nature had not changed. But Velocettes, being idealists,

The ideal-design dreams of Eugene Goodman were translated into a practical design by Charles Udall (above). He says of the LE, modestly; "Given certain terms of reference, the rest became automatic"

went much further and dreamed up an ideal design which history will one day accord the praise it deserved. If there was a fundamental mistake in the reasoning that Ixion made, it was the retention of the motorcycle shape and image. As it turned out, the scooter was nearer to the shape of ideal utility transport in the eyes of the man in the street, by reason of certain advantages inherent in its small wheels. Small wheels which permitted the insurance of a spare wheel and a walk-through platform. Small wheels which ruled out steering to motorcycle standards: but what did Mr. Everyman know about motorcycle standards? In truth, the very instability of scooters made them safe. Riders were not encouraged to take risks.

Velocettes, I say, were idealists and as they went through the list of requirements the design became more and more elaborate. First, they argued, it had to be vibrationless for the comfort and wellbeing of the rider and to reduce wear and tear of the machine. The only motorcycle type engine inherently free of vibration was the flat twin.

"Here, then," wrote George Beresford, who was commissioned to write a why and how book about the LE entitled *The Story of the Velocette*, "came the first big decision to be made and perhaps the most courageous one. Pluckily it was taken and all thoughts of a single cylinder cast aside. Very tempting it must have been to produce a single with its small dimensions and simplicity: but the designers were not to be side-tracked: they took the big view, constituted themselves the implacable enemy of vibration, and came down heavily on the side of the horizontal opposed twin."

So it was all along the line. Fan cooling could, as we have learned from small industrial engines, provide satisfactory cooling under all conditions, but to Velocettes the only proper way to give reliability, long life and silence was the car way, water cooling by water jacket and radiator. Twenty thousand miles without serious attention was the aim. That ruled out chain drive, for a start, even if it had not been ruled out by the adoption of a transverse engine. It was unthinkable to start off with in-line transmission in the engine-gear unit and then compromise to final chain through a right angle drive. For the Velocette people, at least.

And they had a working design already tested in the shape of the Roarer. It only needed scaling down. So did the adjustable rear springing from the Model O. Engine size was dictated by the desire to give the magic figure of 100 mpg with a reasonable top speed of 50 mph - the graphs of the two crossed at 150 cc.

Many of the details which came to be criticized by

Birth of the LE

THE STORY of how P.E. Irving drew, in his words, "the bare bones" of the LE while convalescing at home recovering from, again in his own words, "getting too close to an incendiary bomb" while fire-watching on the factory roof is well known. With Eugene looking over his shoulder, as it were, the two men having agreed on the requirements, silence, smoothness, comfort, weather protection being important, but modest performance acceptable, the result was much more basic than the final design, oddly enough drawn by Charles Udall when he too was convalescing, in his case after appendicitis. Bob Burgess described it pretty well from memory as being an "almost but not quite flat twin", water-cooled but with air-cooled, finned alloy heads. It had, he said, chain final drive, the rigid case supporting the rear wheel on a stub axle, permitting easy removal.

Irving provides a little more information in his autobiography. He describes the engine as "an almost flat twin" which later, long after, when he was writing technical articles, appears to be one of his pet ideas which no one took up. He often said he could not understand why BMW, for instance, did not gain more cornering clearance by "bending" their flat twins up a bit. In this embryo LE he was going to try it out. At no time did he mention air-cooled heads, however, and one of his original drawings which I found among the drawings for the model O which I have, shows water-cooled heads with alternative water connections. The drawing shows a gearbox intended for shaft drive to an underslung worm (abandoned because Hall Green had no machinery for making worm wheels) and a more compact design with final bevel gear to allow chain drive. It was considered for the low speed machine envisaged that a rigid frame would be satisfactory and a sprung seat pillar would provide adequate comfort and save weight and cost. The really significant feature of this first drawing was that *the rider's feet were to be placed ahead of the cylinders allowing a sitting position* as in a kitchen chair, a position chosen by all the successful scooter designs which were to follow.

Clues left in Irving's autobiography about the embryo LE suggests a box or U-section frame and a pressed steel mudguard with the engine-transmission dependant from it. Irving does not mention a fork but Burgess recalls a simple plunger. He states it was intended to use a fork with sprung fork ends like the speedway forks which Messrs. Webbs made at the time. This was a simple design with a small range of movement; versions are still used for speedway and grass track racing. The final LE fork, by comparison, could be said to be over-engineered for its purpose. Final drawings were never completed and no prototype was built to this basic design which, Irving says, was intended to appeal to young women and sedate office workers for use at moderate speeds on urban and city roads.

It may well be that after Irving left the firm, and the project was taken over by Charles Udall, it was decided that the market could be enlarged by making a more competent machine but unfortunately the end product, though an engineering tour de force only comparable with and perhaps exceeding that of Granville Bradshaw's legendary ABC, missed out on the intended market but found a completely unexpected one, the Police.

LE assembly line where the engine, gear and final drive units for the revolutionary everyman two-wheeler were assembled

motorcyclists and were subsequently modified were deliberately adopted to please non-motorcyclists, so the criticisms can hardly be valid in relation to the original conception. Hand starting was provided because non-motorcyclists found kickstarters an abomination (ingeniously, the hand lever was linked to the centre stand so that this was lifted as the engine was started.). Charles Udall invented this mechanism: hand gear-changing and three speeds were adopted for the same reasons.

The clean-to-ride and easy-to-clean requirements were met to a large extent by the generous mudguarding and the built-in legshields. Certainly when a windscreen was fitted — no provision for one was provided, for at the time windscreens had not become commonplace — it was certainly possible to undertake quite serious riding in no more than a raincoat. A touch of luxury was provided by the warm air from the radiator, but looking back, it seems surprising that no attempt was made to shield the rider's hands, body and face. Anyone approaching the problem of an everyman machine from first principles might have thought this essential, but the LE was, of course, designed by life-long motorcyclists who were used to braving the elements. Although I would not have recommended an owner to hose down an LE, as Ixion demanded, it was certainly a machine that could, apart from the wheels, be washed down easily with bucket and sponge.

In the glass-fibre-styled Vogue model there is the dream motorcycle of both yesterday and tomorrow. The kind of machine — the car on two wheels — that enthusiasts once dreamed about and may well do again when today's fashion for ultra-sporting machines fades. If there is anyone truly interested in motorcycle design who is not familiar with the detail design of the LE I earnestly recommend him to obtain a copy of George Beresford's book on the machine and study it thoroughly. Never, I think, in the history of motorcycles has a machine been designed and built with such sincerity and idealism. In time to come it will be held up as a memorial to the greatness of the Goodman family, yet at the moment such examples of the 150cc models as still exist are virtually valueless. And this is the machine which created a sensation when it was unveiled, which by its silence captivated man and women who had previously hated all motorcycles and inspired George Beresford, a motorcycle journalist of great experience, to write, obviously from his heart, "It has that engaging personality which was possessed by its forerunners and which is the outcome of genuineness and sincerity: it reflects the mood and feeling of its originators, it reveals their honesty of purpose. It is small but by no means insignificant, it is proud but not in any way arrogant: it is unique and yet not out of place. Like its makers, it is a leader which manages with complete charm to carry its laurels with becoming modesty."

Has any other machine ever inspired such an eloquent panegyric?

What did Ixion have to say when, in the autumn of his life, he saw his dream come true, fulfilling almost to the letter the requirements he had set out in 1929? His tribute to the LE was to devote the final chapter of his magnum opus, *Motor Cycle Cavalcade*, to its appraisal. He revised his 1929 thesis to include a few additional requirements which the LE had now provided. Absence of vibration, silence, ease of starting, easy cleaning and protection from weather were still listed, but he was now insisting on shaft drive, accessibility of all components demanding periodical attention, the lowest possible weight consistent with the technical requirements, the lowest price consistent with rather an exacting specification, and a good standard of speed and climb in proportion to the engine capacity and, finally, a really difficult requirement which only the LE among lightweights of the period was likely to satisfy... the ability to run 20,000 miles before any major reconditioning such as decarbonization or a rebore.

He went on to suggest how his new list of ideals could be achieved technically, and it is now clear that he was writing about the LE.

"The claims which I advance for this exciting prototype of future designs are based," he wrote, "upon a single undeniable fact. It is entirely proof against all the criticisms which have been advanced against practically every motorcycle in the world, by practically every man and woman who have ever ridden them. In every particular performance (except one) it touches the peak of every motorcycling dream."

This was the opinion in 1950s of the doyen of all motorcyclist journalists. He was not clairvoyant enough to look forward to the present day when so many of his ideals of silence, refinement, weather protection and unobtrusiveness in the eye of the public have been cast into the dust and trampled on by a minority of exhibitionists who have undone in a few years the good work which took him a lifetime.

A man less experienced in the motorcycle world and less familiar with the vagaries of human nature might have gone into raptures over the possibility of every other man and woman in the streets taking to two wheels on LE Velocettes or similar machines. Ixion had no such illusions, made no over-optimistic predictions of its commercial success.

Reading between the lines, I detect instead that Veloce was already assembling reasons for the impending failure of the LE to sell in vast numbers. For though the LE sold very well for a completely revolutionary design, well enough by normal motorcycling manufacturing standards, such a limited success in the case of a machine designed for mass production by a firm staking so much on a single throw was little short of failure.

He ended the chapter, and the book, with a prediction. "I regard it as destined to beget a noble breed of new models many of which will assuredly carry other transfers and nameplates. It is in sober fact one of those revolutionary motor cycles which looked like the original Douglas, Triumph and Scott machines which exerted so marked an influence on design at the now distant dates of their respective debuts."

I suspect that this final chapter gave Ixion one of the most difficult tasks of his long career. For a lifetime he had used his pen and his influence to further the cause of motorcycling and to coax manufacturers to design more advanced and refined motorcycles. In his youth he had jeopardised his health in testing primitive bone shaker veterans, in his later years he had stood fearless as the defender of the faith of motorcycling, dreaming always of a day when motorcycles would be as refined as cars and as socially acceptable. The LE Velocette was his dream

LE 150: Ease of starting is demonstrated at the launch of the machine in 1948

The Goodman brothers, Eugene (left) and Percy, with an early production 150 cc LE and a 1913 two-stroke

machine come true, but the years of experience as a journalist in a field where it was always important to keep on the right side of all advertising manufacturers, even at the risk of sycophancy, prompted a cautious approach. He avoided constant repetition of the title LE by the unsubtle device of referring to it as the "X" machine. His professional conscience thereby appeased, he then went on to extol its virtues to the extent of a few more thousand words.

But though it was so near his ideal machine he had already formed reservations about its chances of commercial success. He could see, for instance, that it would make no appeal to existing riders unless its power was increased but saw it as the prototype of the motorcycle of the future.

Never before had a machine been designed with such uncompromising idealism to provide "Mr. Everyman" with safe, silent, sophisticated two-wheeled transport. Seldom if ever has a motorcycle made a such an impact on the non-motorcycle minded public or obtained at the time of its introduction so much gratuitous publicity in non-technical mass media. Yet in the main Mr. Everyman remained unconvinced that here was the answer to personal transport. The majority of LE 150 Velocettes were, I think, sold to the more elderly motorcyclists who had become too mature, too staid, perhaps too timid any longer to lust for hog buses. True, a percentage of commuters more daring than the rest saw LEs whispering by as they waited in bus queues and bought them but it was still a small percentage of the vast total of potential customers.

The ex-motorcyclists who came back to the fold via the LE, and the newcomers who found motorcycling enjoyable, soon became dissatisfied with the limited performance. It was not so much the lack of maximum speed — the 150 model would wind up to 50 mph and hold it — but the lack of acceleration and lack of reserve to cope with wind and gradient. Mr. Everyman, the mythical figure conjured up by the pundits but seldom met in the flesh, might have been satisfied with the 50 mph and plus 100 mpg performance but those who actually bought LEs were not satisfied for long.

Velocettes, to their credit, took note of the criticisms and a couple of years later enlarged the engine to 192 cc, which made all the difference. At the same time it enabled them to improve certain details which had given trouble in service... the most obvious refinement being the addition of an external oil filter. But although the enlarged engine with its improved performance and stamina pleased the existing LE enthusiast market... already a loyal and sometimes eloquent one-marque cult... it did not of course make any more appeal to Mr. Everyman. So the LE became and still remains a specialized machine for a particularly specialized market. Alas, a market which was inevitably to shrink and disappear completely for it was largely made up of elderly types who had run the gamut of conventional machines and come to appreciate the pleasure of smooth, silent, economical motorcycling.

Where did the LE go wrong? Why did such a dream fade and high hopes crumble to leave behind the bitter ashes of disappointment for those concerned. Until Mr. Honda came to count his miniatures by the million it might have been said that the Everyman market was nothing but a mirage, that it did not exist and never had existed. Unrepentantly I repeat what I have written before. The real mistake, I maintain, was that Velocettes set their standard too high for the market... like offering an exquisite Swiss watch to a navvy. A Rolls-Royce for a Ford market (with no disrespect to a Ford which is and always has been a classic example).

Price was the first snag. The underlying aim had been to market the LE at £100 (perhaps less when it was designed), cutting the cost of the expensive (by two-wheel standards) design by mass production on a scale hitherto unknown in the industry. The factory was completely reorganised on the moving track assembly line system. The capital investment on the production of the LE and the necessary re-equipping of the factory represented one of the greatest expressions of faith the industry has known. But you can only mass produce articles if you can sell them in mass, unless you have unlimited capital resources for stock piling. Velocettes, a small company by comparison with others, needed to recoup their capital expenditure (estimated at £100,000) quickly. The starting price, therefore, had to be economic. Prices could be slashed later when the tooling was paid for. So the original price was £116 plus £31 PT - nearly 50 per cent more than the original aim. The BSA Bantam, which was a winner by comparison, sold for £80 with tax. The price of the LE ruled it out of the running for the hard up ride-to-work artisan. Those who could afford an LE without so much as flinching were sometimes put off by the lack of performance... the lack of putative performance in many cases, for the LE was adequate if not exciting. But a man does not buy a Rolls-Royce because he wishes to drive at 100 mph He does so because he wants the luxury of more power than he needs.

Power to leave the herd behind, not often used perhaps, but exciting to contemplate. The LE was a Rolls-Royce with Austin Seven performance, not a combination which

In 1977 the Chief Engineer, Kent County Constabulary wrote: "My opinion is that the LE was the best lightweight motorcycle we ever had. They outlived the Ariel Leaders, BSA Fleet Stars and Matchless G15s. They were liked by riders and the mechanics who maintained them. The Government should have bought the firm in order to supply all Constabularies with a machine ideal for their purpose."

has ever appealed. It may have been the machine which the Goodmans craved in their motorcycling maturity, the machine which erudite enthusiasts had pleaded for (while continuing to buy conventional machines), but it was not the machine Mr. Everyman could afford, or properly appreciate if he could. He was not to know that it steered like a TT replica, that it had an engine that was a triumph of engineering. From his point of view it was a motorcycle, it would have punctures, it would skid, he would get wet and cold. It looked a bit odd too. A rather unhappy confusion of straight lines and curves. Mr. Everyman, afraid of anything mechanical, must have been alarmed by the exposed pipes and things. Not until scooters arrived with everything enclosed in styled tin-wear, like fridges and washing machines, were his fears assuaged. Of course, the LE should have been completely enclosed and styled with purposeless bulges, needless tin-wear and rust-prone chromium gewgaws. But I hope, after reading this far, you know that Velocettes could never stoop to such chicanery.

It should have been launched with fanfares and starlets and sex symbols and the dealers stocked with them on a sale or return basis.

Instead, of course, through material shortages and a few teething troubles (no pun intended but the primary reduction gears were noisy and finally had to be lapped in on a special rig, and the first big ends expired prematurely), the flood necessary to cash in on the first wave of enthusiasm was but a trickle. And Velocette dealers, mostly small, dedicated motorcycle enthusiasts, were not psychologically or often geographically best placed for selling a new product to a new market. I will say this for them, though — they did, more out of loyalty to the firm than from pecuniary gain, try very hard to sell the LE. Whatever chance the LE had of finally making the grade in the popular utility market was finally torpedoed by the arrival of foreign scooters. By comparison with the LE they didn't steer, didn't stop, weren't vibrationless, not very quiet, and didn't go long distances without attention. But they had style and colour, little wheels you could change for a spare and the works were out of sight and presumably out of mind. They were publicised by starlets in shorts, became sociably accepted because they were not motorcycles, and Mr. Everyman and his wife bought them like they bought washing machines — on stylized appearances.

The ideal design dreams of Eugene Goodman were translated into a practical design by Charles Udall. He did the work at home while recuperating from appendicitis.

"Mr. Eugene thought it would be a wonderful opportunity," recalls Udall wryly. Modestly he added: "Given certain terms of reference (the everyman machine requirements), the rest became automatic."

This is much too modest an assessment of a design which for its ingenuity, compactness and overall fitness for a specialized purpose will eventually go down in two-wheeler history as a masterpiece.

Why, I wondered, was it called the LE? Was there any subtle reason for the initials which slipped so easily into common usage? Apparently not. It just so happened that when drawings for the machine began to go through they had to be given some identification to avoid confusion with various Government contract jobs. So the engine unit drawings were headed "Light Engine" which was sufficiently non-committal to prevent the secret leaking out, and at the same time disarmingly accurate. Light Engine became abbreviated to LE and the designation stuck. In a more image-minded era it might have been called The Leader and then Ariels would have had to think of another name.

Velocettes sprang a real surprise at the 1956 Earls Court Show by introducing the transverse flat-twin ohv 192 cc Valiant. This was a typically Hall Green attempt to meet the growing demand for an under-250 cc sports machine, not with a rehash of the old MOV single (which the trade asked for and might in the end have been a more commercial success) but with a very sophisticated design of the type which armchair designers have favoured since the demise of the 1920 transverse ABC and connoisseurs have since revered in BMW guise. In all respects, save perhaps that of silence, the Valiant could be regarded as a miniature BMW. Such an advanced layout would have been out of the question for most manufacturers (though similar layouts had appeared briefly in Germany and Japan) because the tooling costs would have been prohibitive but it was within the bounds of possibility for Velocettes who had the foundation in the LE. All they had to do was design ohv air-cooled barrels and heads to fit the LE crankcase, and they had the engine, gearbox and shaft drive all ready. A duplex-cradle tubular frame to suit was easy... it was very like a miniature version of the one sketched out for the racing four. Even the LE forks were ideally suited.

There was, of course, a little more to it than that. The crankshaft assembly had to be beefed up to take the extra punch and revs, and four speeds were considered necessary on a high-performance lightweight like this. The cylinder heads were a particularly nice piece of work. Cast in light alloy with shrunk-in valve inserts, they were of hemispherical combustion chamber shape. The valves were of generous size for the capacity of the engine, were closed by single coil springs and operated by rockers which pivoted on rocker standards cast in with the head. Adjustment was by the well tried Velocette method of eccentric rocker spindle as on the Mark 2 KSS and the KTT. The ohv mechanism was enclosed by oval saucer covers reminiscent of a BMW. Pushrods were in light alloy and operated by piston-type cam followers, as on the LE. The same car-type lubrication system was employed, with external feeds to the valve gear and external pipes to the sump. The gearbox was modified by adding an extra gear to the cluster and a normal Velocette positive-stop-operated rotary cam plate selector mechanism was added, being mounted face down in the gearbox lid. The heavier crankshaft assembly and footchange gearbox was employed on the later version of the LE.

When the Valiant was first announced, a single Amal Monobloc carburettor was fitted completely enclosed in a pressed steel or a glass-fibre nacelle above the crankcase. This nacelle was a trifle futuristic. With the electric horn grill set in the nose it looked rather like a miniature jet engine. I criticized it at the time, though this was a period when fashion was all for enclosure and ovoid shapes, and I still do. Some illogical instinct puts me off any attempt to cover up a bikes works by adding "falsies".

Soon after it got into production, two carburettors mounted direct on the heads were substituted, with a balance pipe between them. The big ends tended to be a touch fragile, so a workshop bulletin soon winged its way to all Veloce dealers; enclosed with it was some fairly thick compression plates that they had to surreptitiously fit to the machines when they came in for service. That is why some owners found their machines not as sparkling *after* a service! One point always puzzled me. Suspension was by vertically disposed Woodhead Monroe hydraulic units and there was no adjustment for load.

On the road the Valiant lived up to expectations. It was very smooth when revving, so smooth that the makers had to stipulate a rev limit of 7,000 rpm in the gears because it was all too easy to let it buzz on and on to the point of no return. I recall it being a little harsh and lumpy at low revs... there was the old flat-twin couple effect and it was difficult to get the two carburettors spot on. Steering, braking and suspension were of the race-bred order expected of a Velocette. It was difficult to realize that the engine was under 200 cc, for with its 70 mph maximum and 60 mph cruising it behaved more like a 250. Altogether a delightful machine which never received the credit it deserved because so few were sold. The reason was undoubtedly the price. It was announced at £181, including £35 purchase tax, and by 1957 it had risen to a few shillings under £200. A 197 Villiers-engined Francis Barnett cost £160, a 199 cc Triumph Tiger Cub £143.

Sophisticated design: the 192 cc Valiant

In 1963, 15 years after the launch of the LE , a glass-fibre-bodied version, the Vogue, made an appearance. It was not a success

The Velocette motorcycle was not so much a product of commerce as a proliferation of the kind of motorcycle that the Goodman family liked. I like to think that if the family had been blessed with limitless wealth they would still have made motorcycles for the pleasure they gained from designing, developing and manufacturing them and for the pleasure they could give to those who used them. Not just the Goodmans, of course, but the scores of loyal workers who spent their working lives at Hall Green giving of their willing best and enjoying a pride in their work which demanded that every Velocette which left the works, be it racer or humble two-stroke, was as right as they could make it. Determined that *their* bikes should be that little bit better than those from the other encampments at Selly Oak, Armoury Road, and Bracebridge Street — a sentiment which, of course, was shared by their rivals at those factories, for I am thinking about the days between the wars. The days before motorcycles became units in rather abstract production figures; before they came to be classed as "consumers' durables"; and when each model, if not each individual machine, had a distinct personalty which was an extension of the personalities of those who had designed and built it. When "quality control" had not been thought of as a term or as a necessity but was the natural outcome of pride in workmanship. Double checked when, as often happened, one of the "bosses" would snaffle a bike from the production line and ride it home to lunch or away for the weekend.

Irving's Light Engine design: more revolutionary than it appears

LOOKING at the drawing (reproduced here) is like being given half the pieces in a jigsaw with no sight of the completed game. Though I could get the general idea there were many questions beyond me. Only a draughtsman can read another draughtsman's drawings and essay to interpret the designer's intentions. I took the drawing to Bob "Vulcan man" Higgs. We soon agreed that the long gearbox was intended for shaft drive where its length was no disadvantage, possibly an advantage because it shortened the drive shaft; the short box was particularly short to make room for a bevel to turn the drive to final chain. The long box gears locked on to a hollow main shaft by balls pushed out by an internal selector tube (a method popular on Italian lightweights years later). I left the drawing with Bob for a few days. His conclusions were surprising.

"Although at first sight it looks similar to the final LE engine/transmission unit it is in fact quite different and very, very clever. In many ways it follows mass produced car practice having a monobloc cylinder and crankcase casting, presumably in alloy, to which is bolted another casting which is the top half of the gearbox. They share a pressed sump like a car so the clutch is wet and no seals are necessary internally. The crankshaft runs in plain bearings, most likely clamped by a bearing cap from underneath like a car crankshaft. The short three-speed gearbox is very unusual. Bearings for both shafts are carried in saddles bridging the upper gearbox casting so the gear cluster could be fitted or removed complete. Space is saved by having moving dogs on both shafts and running one lay shaft actually on the boss of a sliding dog rather than on the side. The selector forks would be operated by a selector mechanism on top of the box, probably a bolted-on lid like a car. No starting mechanism is drawn but a quadrant mounted on the gear-

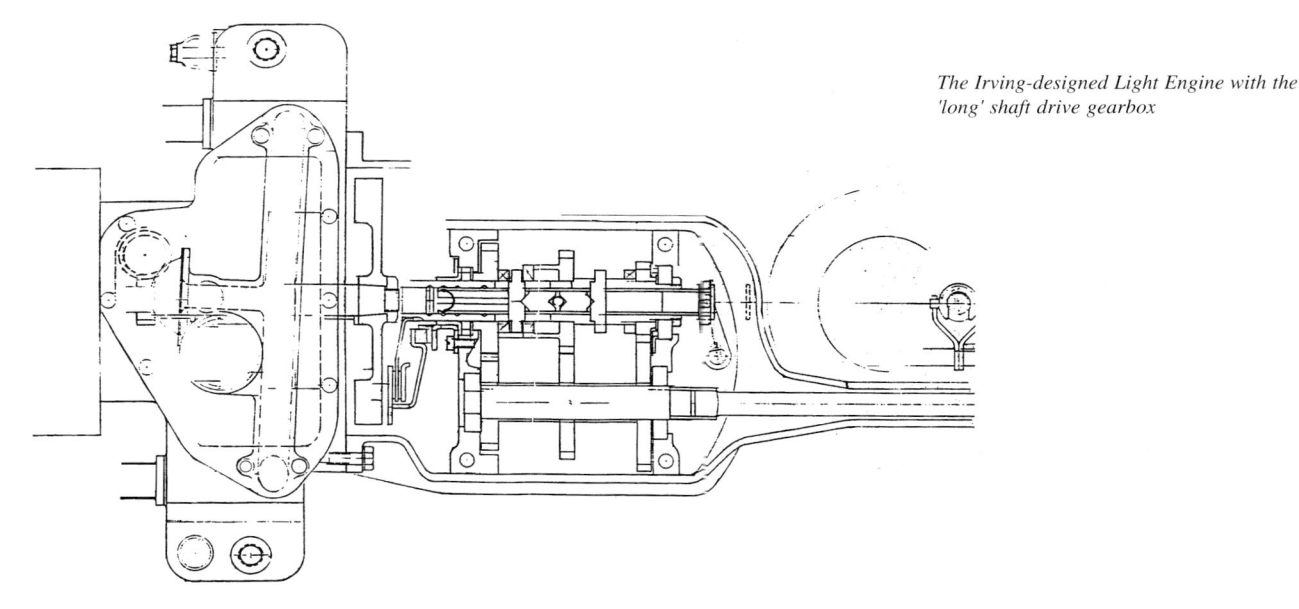

The Irving-designed Light Engine with the 'long' shaft drive gearbox

The 'short' chain-drive gearbox design

box lid could have engaged the fixed pinion on the end of the gearbox main shaft.

When you study the engine drawing it gets more and more interesting. The included angle of the cylinder part of the monobloc is 140° which may not have made such difference to the balance but makes a lot of difference to the width of the crank case.

Angling the cylinders upwards gets them out of the way of the crankcase sump joint faces and there being no cylinder base joint, I imagine, the saving is 2in at least. The open top of the cylinder/crankcase casting is closed by an alloy casting which includes the induction system and the float chamber and mixing chamber of an Amal type carburettor, all kept warm by the engine. This multi purpose casting also covers the valve stems and tappets and would allow access to the cam shaft and crankshaft. The crank webs appear to be split to secure the crank pins like the early LE. The cylinder bores are offset by 1in to allow adequate width for the big end bearings, but the valves are offset in the blocks so they are not offset so much side to side and I can only imagine this was done to enable a two-lobed cam to operate the four valves... a cost saving not to be ignored. Offsetting the valves left insufficient room for a vertical water outlet on the offside hence the bolted on adaptor. No camshaft is drawn, nor a drive. There is what appears to be a skew driven vertical shaft to the front which most likely drove an oil pump in the sump and could have driven the camshaft through another skew gear train. There does not seem room anywhere else."

One cannot but admire the original ideas which went into this design which would, I am sure, have been much easier and therefore cheaper to produce than the final design. Unconventional by motorcycle standards, it followed methods of construction developed in the car world for mass production and low-cost manufacture of engines and transmissions which had to give long life with smoothness and reliability but did not have to look pretty or imposing as did motorcycle engines at the time. The requirements for an 'Everyman' power unit were the same as for a car but on a smaller scale. It was therefore logical for Irving to adopt car practice with some neat detail touches of his own. What a pity that the exigencies of war time service under the direction of the Ministry of Supply took him off the job and sent him to Associated Motor Cycles to work with Joe Craig on, believe it or not, the embryo Porcupine. I suppose it came under the heading of 'preparing for peacetime.'

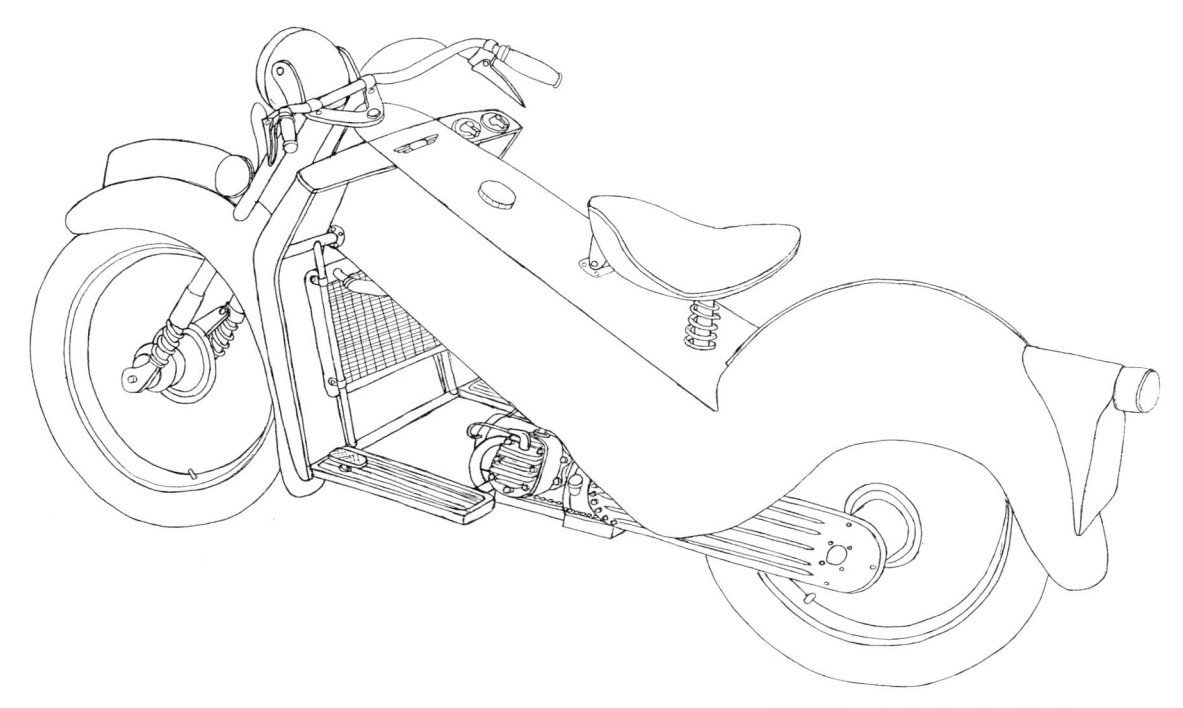

Could this have been what Phil Irving's version of the LE would have looked like? Drawing by Jonathan Wortley

Arthur Taylor

DEALERS in the early days were personalities known intimately to the manufacturer, and the bond was very close in the case of Velocettes. Such euphemisms as "retail outlets" were unknown at Hall Green.

I thought on these lines after talking many years ago to a typical Velocette agent who over the years became one of the greater Velocette family — Arthur Taylor, of Shipston-on-Stour, a sleepy Cotswold market town. To followers of production races he was a smallish, lean figure with a cap and an inevitable pipe, hovering over or working on a Venom and Viper. A bit earlier he was there in the background when Cecil Sandford (later his son-in-law) was making a meteoric climb to world champion status on lightweights. That climb started with a surprising 250 ohv MOV Velocette which became quite a legend, and that was the real reason why I went to Shipston.

Starting just after the first world war with a little workshop and a solitary petrol pump, Arthur Taylor became a motorcycle agent almost by accident.

"A gentleman pulled up for petrol one day with a beautiful Alvis and asked if anyone round about knew anything about motorcycles. I said I knew a bit — I rode an old Bradbury.

"It turned out he was Norman Downs, Managing Director of New Imperials, and he thought I ought to be able to sell them if I had one in stock. I asked how much one would cost me — I think it was £40. I couldn't afford anything like that and it finished up with me paying half down and the rest when I sold it."

This story will seem altogether too much of a fairy story with fairy godfather and all to younger readers, but this kind of thing did happen in the early 20s when manufacturers would adventure forth and sign up wayside agents.

Arthur Taylor's association with Velocettes was equally typical of the period.

"A local clergyman called one day and said he was thinking of getting a motorcycle and had been told by a colleague that he should get a Velocette because they were very good machines. Could I get him one? I said I could and I did. I had to fetch it from the works which was in Victoria Road, Aston — it was a rabbit warren of a place, all doors and passages. I saw John Goodman and he took me all round and showed me the new overhead camshaft engine they were making and explained why it was so much better than anything else."

The clergyman's Velocette, one of the early two-stroke models, was a very good machine and from this point on Arthur Taylor began to sell Velocettes as well as New Imperials.

I was surprised to learn that in the early days he had tried to repair, though not with much success, one of the 1913 ohiv unit-construction models and had been told later by the Goodmans that it was the only one and had been sold to a relation. I have always wondered why only one example of this highly ingenious model has turned up.

It was not until after the second world war that Arthur Taylor started tuning push-rod Velocettes, and that was almost by accident. Just after the war he bought himself a new girder fork MOV with the idea of riding in trials. With small success, he admits. When Cecil Sandford's father, an old friend, consulted him about buying a racing bike for the lad, Arthur, a great believer in walking before running, wouldn't hear of it.

"Let's see if the lad can ride first,"

With a rev limit imposed by the makers, the Valiant, with its LE based engine, was no racer. Nevertheless Arthur Taylor, age 55½, could not resist sprinting it

he advised, and proceeded to tune up young Cecil's 250 Triumph, on which he went quite well, and then turned his own MOV into a scrambler, on which he went even better. Those were the days before specialized spring-frame moto-crossers, of course.

When Cecil was considered capable of having a go at the "hard stuff", Arthur turned the MOV into a road-racer. It began in a humble way but the bug caught him and it became a hobby for him and his mechanics. I will just mention that eventually the MOV became a force to reckon with in its class at most of the circuits in the early '50s.

It ran in the TT on one occasion and its swan song was at Thruxton where, Cecil Sandford up, it ran away from the work's double-knocker 250 Velocettes.

From a very well-thumbed little notebook Arthur Taylor recalled how they made the MOV fly. Little was done to the bicycle, which had already been fitted with Dowty tele forks, but a great deal was done to the motor. A bronze head was acquired (one of three made experimentally) and an alloy-finned barrel was made from an Alfin blank. A special Martlett piston was obtained which gave a compression ratio of around 10:1 for petrol benzole. The inlet port was opened to 1 1/16in (later it was opened to 1 5/32in but this proved too much) and a 1 9/16in inlet valve fitted, which is the largest size which can be persuaded into a MOV head. Many experiments were made with valve timing — they fill pages in the notebook — but nothing made very much improvement on the standard MOV cam.

In April 1949, it was raced for the first time, at Haddenham, and Sandford finished ninth in the 250 class. It was then tested on the Velocette brake, and turned out 19.3 bhp at 6.500 rpm, falling off to 18.6 bhp at 7,000. Valve float set in at 7,300 rpm.

Eugene Goodman got interested in it and married up a Mk VIII cam to the MOV cam wheel, but this gave too much lift and the valves tangled.

In June 1949, Les Higgins rode it in the Lightweight TT. It lapped at between 32 and 33 minutes and held 6,900 rpm down the Sulby Straight, proving faster than Excelsior Manxmen but slower than the 250 Rudges. When Higgins came off at Keppel Gate it was lying eighth and was the third British 250. Sandford rode it into fourth place at Anstey and won both 250 races at Haddenham. Sandford did so well with it that he was given work's rides by Velocette and was later signed up by MV, for whom he won the 1952 Lightweight (125) TT.

In its final development it had a light-alloy head (a post-war MAC head with the valve enclosure box sawn off and filed flat to take the original valve gear). Because the enclosed coil springs became cooked, hairpin springs were then fitted. It was never timed at maximum speed but Arthur Wheeler, who tussled with it on his Guzzi, reckoned it was capable of between 105 and 107 mph

The swan song mentioned before was at the international meeting at Thruxton in 1951. Sandford was entered along with Bob Foster (who worked for Arthur Taylor pre-war) and Bill Lomas on the rather hush-hush dohc two fifties, which were cut down Mk. 8-type engines with valve gear based on the 1936 one-off design. They had not been too successful and Sandford persuaded Arthur Taylor to take the MOV along "in case". After practice, Sandford said he was sure he could go faster on the MOV.

"I said Bertie [Bertram Goodman who was then in charge of racing] wouldn't like it."

"Cecil went along to see him, and he didn't, but Cecil argued that he was used to a rigid frame, and with its lighter weight and better pick-up from the corners it would be better. It was. He shot away and they didn't see which way he had gone. I slowed him down after a bit and Bertie was afraid I had overdone it but Cecil understood my signals and when Bob Foster got within sight of him he just went away again."

MOV 250 cc, converted to racer spec by A.R. Taylor Garages, in the trim in which it was ridden in the 1949 Lightweight TT by Les Higgins

Arthur Taylor did not race in post war years, but enjoyed sprinting. The Viper took him to first place in a 350 cc non-racing class

Altogether the surprising MOV, a machine which Velocettes only looked on as a touring machine, must have been a bit hard for Bertie Goodman to take. Especially when he once came upon Arthur Taylor welding up a hole in the piston!

"You can't do that," he said in horror as Arthur plied his torch. " I've got to, I can't get another piston in time," Arthur retorted. Bertie was equally horror-struck when he learned that the MOV still had its original con-rod!

In the development of the MOV he had considerable help from Bob Burgess, who at that time had left the Velocette works to work for him. "He was a great help because he knew all the bits and pieces that had ever been made."

Of course it was a typical case of horses for courses. The race was run in pouring rain, which would have made it difficult to get the best out of the "peaky" narrow power band dohc motors, whereas the featherweight MOV with power all the way would be much more rideable. And Sandford had already had two wins that year at Thruxton on the MOV, first from Roland Pike at 69.27 and later from Fron Purslow on a cut-down Norton at 69 mph In the rain his race speed was 65.37 and his best lap 67.18 mph

Before this and as part of his racing education, Sandford had ridden a roadster Mk 2 KSS into fifth place in the 1949 Junior Clubman's TT.

His father had been willing to buy him a pukka racer but Arthur Taylor, cautious as ever, said: "Buy him a KSS and then if he's no good at racing he will have a nice bike to ride on the road." He took him over to the Island and introduced him to an old friend living there, Chris Stead, who had ridden Cottons in the TT in the early 20s. Stead gave the young Sandford some very good advice which is well worth repeating.

"The TT course," he said, " is really only a foot wide. If you are not on this bit of road a foot wide, you will either kill yourself or you will not be going fast enough to win."

Sandford passed his first Isle of Man test with honours and was then "allowed" to graduate to a Mk. VIII KTT bought from L. Stevens of London, because Velocettes could not supply one at the time, and ridden home on the road! On it he finished fifth in the 1949 Junior Manx, but was out of luck in the 1950 Junior and Senior TT races, having to drop out with minor troubles.

When Sandford went professional and became a double world champion., Arthur Taylor was "out of work" as far as tuning Velocettes went but he could not leave it alone for long and prepared a Viper and Venom for production racing, machines which many may recall were ridden with distinction by Ron Langston, John Righton, Rex Avery and Eddie Dow. He even had a mild go at sprinting with a Viper and, of all things, a 200 cc Valiant.

Arthur Taylor manufactured a crafty rocker return spring device for upping the rev limit of push-rod Velos. It consists of a coil tension spring about the size of the average stop lamp spring which is inserted inside the hollow rocker and hooked on a plug blanking off the nearside. The spring is wound up to give a torsion effect and hooked to the offside rocker box cover.

Talking with Bertie Goodman

Reminiscences of the one-time Managing Director of Veloce Ltd.

Titch Allen wrote this in the first place as part of his Velo Saga running in Motorcycle Sport in 1969/70; in the interests of authenticity, 'period flavour' and general appropriateness it is reproduced here virtually unaltered from original form.

"Be sure to wear your helmet," cautions the mother as the teenage son of the house prepares to set out on the purposeful Thruxton special (a race-bred cammy motor snuggled neatly into its frame). "Of course," he replies, a little hurt, as if he would never ever think of going out without it — not even to the post box! The Velo thuds into life and the father winces slightly as the revs soar upwards in the gears, the boom turning into a crisp snarl as the overlap timing of a racing cam makes a mockery of the filleted silencer. Frowning, the father talks about the problems of begetting motorcycling sons obsessed with speed and racing images and all the paraphernalia of clip-ons and back sets and racing seats and "bacon slicers" (phoney cooling discs for brakes). But the diatribe lacks conviction and real honesty for he has just told me with nostalgic relish how in his teens he was pinched for speeding and dangerous riding (he pleaded then that the machine would not do the speed alleged, and nor would the standard version, but he had already told me that it had been tuned...).

A typical scene in the life of a typically motorcycling household.

Not, however, the scene you would expect to find at the country house of a pillar of the motorcycle industry, the managing director of one of the oldest firms in the industry. The title may conjure up pictures of Bentleys and butlers and motorcyclists banned to the tradesman's entrance. You could not be more wrong in the case of Bertram J. Goodman, Managing Director of Veloce Ltd., past president of the Manufacturer's Association and past President of the BMF, "Mr. Bertram" in the slightly feudal atmosphere which still prevails at Hall Green, "Bertie" always and forever among his motorcycling friends throughout the world. One-time international racer with a unique record (of which more later), world-record breaker (jointly with others) and tester extraordinary, B.J. Goodman is a real motorcyclist of the old school.

I had ridden down to Worcestershire at his invitation because his days at the works are too full to chat. I found the hamlet with the improbable name all right but missed the house because I was looking for an impressive pile becoming a tycoon — which it isn't and he isn't. Chilled by an early-morning wintry ride, I was warmed instantly by the atmosphere of a house which is used to frozen motorcyclists and big boots dumped in the hall.

The exercise was intended to be "this is your life" research. It tuned out to be a marathon swap of motorcycle yarns - with Mrs. Goodman in the role of the understanding, uncomplaining (much) motorcyclist's wife, adding her experiences which are typical of understanding motorcyclist's wives everywhere...

"Remember the time you took me to a plush ball in a sidecar and then turned it over on the way back?"

"And how you used to bring riders home at all hours of the night and I had to cook meals for them."

This was in the days when Bertie was running the racing team and it seems Mrs. Goodman did eventually extract a promise that he would let her know when he was coming home. Arriving at midnight after a racing sortie with the late David Whitworth, he assured her that he had sent a postcard. He had. Popped it in the pillar box at the end of the road a moment before!

B.J. Goodman learned the importance of cutting wind resistance. Here he tucks away his by no means sylph-like form when testing at MIRA

Director who liked his own bikes

The Vintage Club's TT Rally each year ensures a fascinating historical backcloth to the main business, and this candid camera shot is of topical interest... and there is an amusing story behind it too.

The machine is a beautifully original KTT, the property of Eric Thompson, the Vintage Club's then Secretary, who in stormcoat and beret has just done a lap of the old TT course as part of the Vintage Rally. Pulling on his gloves, his trousers tucked into his socks, and unable to hide the look of anticipation on his face is Bertie Goodman, Velocette managing director, who has jumped at the opportunity of a gallop. Soon he was motoring gaily along with his coat tails flapping — and this is where the fate which looks after vintage enthusiasts took a hand. The PA commentator, Harold Rowell I think it was, drew the crowd's attention to the fact that the Veloce head man was letting his hair down, and added as an afterthought that he knew,` but Bertie didn't, that lying dormant in the Island was just such a KTT machine. As a result Bertie Goodman followed up the tip, bought the machine, and with the help of Eric Thompson and the Vintage Club restored it to original mint condition.

Fortunate indeed is a motorcyclist who wins a wife like that, a wife who is a motorcyclist herself.

And so, Bertram J. Goodman, "This is your life"!

I introduce my victim in, I hope, authentic TIYL style... and it is an illuminating chapter of the Velocette Saga.

He is, of course, the son of Percy Goodman, elder son of John Goodman, founder of the Velocette dynasty. His father was the creator of the immortal overhead-camshaft Velocette which did so much to establish British prestige throughout the world. It may be argued that Nortons won more races between the wars, but the Velocette contribution was to produce over-the-counter racers which all could buy and race with equal chance of success.

"My first recollection of motorcycles," said Bertie, "was that my father used my bedroom as a drawing office... I realize now that he must have been working on the first cammy. I remember asking him details about engines — to avoid being sent to bed — and him explaining to me the principle of the Otto cycle. He never pushed motorcycles at me but he was always willing to spend a lot of time on the answers to my questions. Many a time he would come home on a motorcycle, and I remember when he first took me for a ride sitting on the tank... I had difficulty in getting my breath. When the 30 mph speed limit came in I remember him saying that he was noticing the houses on the way home for the first time.

"To the locals he was always 'that mad man on a motorbike.' I first learned to ride when I was nine. A works tester taught me to ride a GTP in a field which is now our car park. I was given a GTP when I was 12 or 13 and used to push it to a playing field where I gave other children rides on the back. One of them was Joan Hazlewood, who now produces the BSA house magazine. The local council eventually stopped me riding on the playing field. As soon as I was old enough to get a licence I was given an MOV. It was the only machine I ever really owned myself. I soon hotted it up, of course... the cam followers were solid in those days and my father told me to drill a hole in the web."

When he was "pinched" for speeding the police alleged that he was doing 72 mph with a pillion passenger.

After this he was "grounded" by his father and given a car. He visited the TT and the Ulster and was fired with the ambition to race motorcycles, but his father was against it, though he did relent enough to let him have a go on a KTT at Donington during tests of the Waycott ISDT outfit. He was supposed to be under the supervision — note how I avoid the obvious pun — of Billy Wing, regular Velocette performer at Donington. Just as his cousin Peter, with the impetuousness of youth, had showed Billy the way round, Bertie set out to do the same. Neck and neck down the Melbourne Straight they went, leaving their braking for the corner much too late. Bertie made it. Billy fell off, saying afterwards that he had been trying to get in front to slow his pupil down.

On leaving Birmingham University Bertie was, like his cousin, apprenticed to Alfred Herbert Ltd., the tool-making firm at Coventry. While there he became interested in the Model O on which development had been halted by the outbreak of war. He rebuilt it and used it for a 40-mile-a-day commuting trip.

"A very stable machine," he recalls, adding: "of course, I usually got up late and had to hurry. One morning I thought the steering seemed rather light, but I thought no more of it until I passed a motorcyclist crawling along trailing his feet. The road was filmed with ice!"

His apprenticeship completed, he volunteered for the Fleet Air Arm and until his papers came through worked at Veloce Ltd. While courting the girl who became his wife, he used the Waycott outfit and there was as memorable holiday trip to Wales with his fiancée and his father. A busman's holiday, of course, being the Goodman family. They were testing the prototype LE, spending day after day rushing it up and down Welsh mountain tracks and quarry workings. Trying to break it. The Waycott outfit had to go through the same mill, of course, and in it, a protesting passenger.

"I told his father in no uncertain terms," said Mrs. Goodman with a chuckle, "that if this is his idea of a holiday

A managing director who did not mind getting his hands dirty. Bertie Goodman changes a sprocket between performance tests

I was going home."

But of course she stuck it out and later became an LE enthusiast.

Talking about the Waycott outfit reminded Bertie of his first experience with a sidecar outfit. Seems he agreed to go on a sidecar camping tour of Norway in 1937 with a non-riding friend. Never having ridden an outfit before, he thought he had better get some practice. He did not get far down the road before the sidecar took charge and there was a phenomenal avoidance.

On the MSS outfit they covered 3,000 miles in 10 days on a total of £10 and arrived back with 10s left. Happy days. Demobbed, he went over to watch his cousin Peter ride in the 1946 Manx (he finished sixth in the 500 cc race) and next year entered for the Junior TT himself.

"Was not this a bit of a nerve — making the TT your first race?" I queried.

"Not really. Things were a bit different then. Racing had only just got going and I went over as a private owner taking the Model O as a hack. I did 54 laps on it learning the course and I had no practice problems when I came to take out the KTT. Mind you, I didn't take much notice of the chaps who told me how they went flat out down Bray Hill. I was only riding for fun. But I made the mistake of having new tyres fitted just before the race without an opportunity to give them a run. At Laurel Bank the bike was all over the place. I didn't realise the back tyre was going flat. I pressed on knowing that my uncle would be watching at Birkins, but it was hopeless. By the time I toured back to the start it was dead flat. Just as well I did not blow it up and carry on for it turned out to be a faulty valve cap. Strange, but Ken Bills had the same thing happen at the identical spot."

Third place in the 1947 Ulster was some consolation. An uneventful ride. Came the 1948 Ulster — by now he was running the firm's racing department — and his jinx struck again. He decided at the last minute that the high wind demanded a change of sprocket. Readjusting the rear chain he found that both rear-wheel bearing adjusting nuts had split. He was out before he started.

See what I mean about unique international racing record? I can think of no one who had such appalling luck and yet finished third in a classic in his second race. After that he dropped out of racing, though he entered the TT as a reserve so that he could try the bikes out under racing conditions. The 1949 Ulster was his final indulgence — being at the end of the season he could relax — and it was the only time his wife came to watch. Sure enough he fell

Bertie Goodman... third in the 1947 Ulster 350 GP

off, bending the bars, when the exhaust pipe grounded.

"If I had carried on, and I suppose I could have done, we would have won the team award," he said.

The late 40s and early 50s were hectic for Bertie Goodman. Velocettes were not officially sponsoring racing but they were producing KTT models and helping private entrants. As head of the racing department he worked all hours to produce the machines at a time of desperate post-war shortages and then rushed all over the Continent with the "Circus" to keep them "flying". His wife did not see him for weeks at a time.

In 1948 three well-to-do enthusiasts came up with a really ambitious project. Dennis Mansell, formerly a director of Nortons, and Nigel Spring, one-time rider and pre-war sponsor, first got together. They would sponsor Freddie Frith and Ken Bills in the classics if Velocettes would build some "works" KTTs. Dick Wilkins then offered to sponsor Bob Foster on the same terms. This was a chance to "Back Britain", and Velocettes co-operated wholeheartedly, extending their help to David Whitworth who had shone brilliantly pre-war and was hammering bravely round the Continental circuits as a privateer. The first post-war "semi-works" machines were hotted-up versions of the pre-war mounts using the big fin "works" heads and barrels, a bigger inlet valve (later to become the standard Venom valve) and 1 5/32 RN carburettors. There was an auxiliary scavenge pump as used on work's machines before the war. A slim gear pump built into the bevel housing cover and driven like the standard pump from the half-time gear shaft to evacuate the lower bevel box which normally drained into the cam box drain passage in the offside crankcase half. A modified front brake had a cast-iron drum bolted to an electron hub; it was less prone to expansion than the original cast-in design.

The 500 mounts used by the syndicate were substantially pre-war models. With one of these machines Frith won the 1948 Junior TT and the Junior Ulster.

During the winter of 1948 Bertie Goodman designed a double-overhead-camshaft version of the 350 and 500 cc models.

In effect he resurrected the abortive dohc design of 1936 — based on a "dog-kennel" engine with semi-detached cam box — and adapted it to the all-enclosed engine. The cam box was reshaped with a flat base to bolt directly on to the alloy head which had its rocker box portion sliced off, leaving only the rectangular troughs for the hairpin springs. Metal was added to the head casting to provide a firm base for the cam box and a foundation for the numerous small holding-down bolts. Electron was used for the cam box, crankcase and timing cover.

"I used a lot of midnight oil," said Bertie, "often worked all night and all weekend. There's no place as lonely as a factory at the weekend and I remember once I heard footsteps in the early hours. My hair stood on end. It turned out to be the men coming in to light the boilers! We built six 350s and two 500s and the first engine came off the bench the day TT practice began. Men who had worked all day, then worked all night to fit them into bicycles."

There had been a lot of minor development problems with the dohc engines. As in 1936, the standard Oldham coupling for the vertical shaft objected to the extra load. It was strengthened by fitting a collar round the slot and reshaping the coupling. The shaft itself was waisted to make it more flexible in torsion. Work's engines were by now using roller races in the bevel shaft housings, and a double-row self-aligning race was used on the drive side of the 500 model. Engine revs in the 8,000 region were possible with the dohc design and led to rapid wear of cams and tappets. It was overcome by welding on a hard skin of Delchrome, but the problem then was that the necessary small oil ways could not be drilled. The answer was delightfully simple. A small pipe was attached to the part and as the operator ran the molten metal over the surface he blew lightly down the pipe. A "puff" at the right moment and a blow hole appeared where the oil way should be.

The dohc Velos were not, of course, ready in time for the traditional try-out of work's TT machines, the North West 200. To get some training, Freddie Frith and Ken Bills, the Mansell-Spring syndicate teamsters, took a couple of "off the shelf" standard KTTs over to Floreffe where they found that the new 7R AJS ridden by Les Graham was really flying. So, too, incidentally, was a 250 Benelli ridden in the 350 race by Ambrosini. It took Frith five laps before he could get past it, but he could not catch Graham, Bills was fourth. They rode the KTTs back to the port, being picked up by the police as they pobbled, sans silencers or any legalities, through Brussels. A spot of dead pan "no speaka de language" got them out of that one, and they didn't risk it with our police but took the bikes up to Liverpool on the way to the North-West 200 by van. In Ireland they rode them over to Portrush as a matter of course. The Irish folk would have been disappointed if they had not. Although the Velos were absolutely standard and the AJS and Norton machines were the work's jobs for the TT, the Velo men thought they might get a slight advantage by avoiding a fill-up.

Fitting a couple of oversize tank, they carried out capacity and consumption tests... were busy with funnels and measures when Harold Daniell and Artie Bell, their Norton rivals, walked by. "It's no good," said Harold drily, "we've tried it!"

Off-the-shelf Velo or not, Frith gave the work's bikes a real run for their money. All the way he was dicing with Bill Doran on the AJS and Daniell, each one of them holding the lead at one time. Near the end Doran went out with gearbox trouble and Daniell crossed the line a mere wheel or so ahead of Frith. So shattered was the man with the chequered flag that he let them both go by to do another lap! Both Daniell and Frith completed the race on a tankful! Bills, unfortunately, was involved in a mix-up and broke his collar bone.

Study in style (1): the incomparable Freddie Frith on his way to victory in the 1949 Junior TT

Study in style (2): Ernie Lyons takes third place in the 1949 Senior TT (earlier in the week he managed second in the Junior)

It was now obvious that it was no longer merely a struggle between Norton and Velo; the AJS was a formidable contender, and so it turned out in the Junior TT.

On the first lap Graham led, closely followed by Doran, with Frith and Daniell third and fourth. Graham went out with clutch trouble, but Doran hung on to the lead until the last lap with Frith some 10 seconds behind on time, though first on the road because of his 1948 win. When Frith finished and waited in suspense, Doran was at Ramsey. A last-minute spurt and he could win. But at the Gooseneck his gearbox, the weak link in the new design, failed. Ernie Lyons, who had taken over Bills' machine, did a wonderful job. Taking on his friend Artie Bell, the Norton leader, he finally beat him to finish second.

It was wonderful debut for the dohc Velo, a final vindication of the original 1936 design and a fitting reward for the months of hard work and midnight oil put into it by Bertie Goodman and his loyal workers.

Nigel Spring revealed afterwards that Frith was riding to signals. Given the "flat-out" at the start of the last lap, he had by Sulby turned an 11 seconds deficit into a nine second's profit. Frith confessed that it had been a hard race. Observers' reports were that he was really trying, and he had been troubled with gears jumping out when changing from second to third. The cause was apparent when the machine was inspected. The gearbox had pulled back and the primary chain was as taut as a bow string.

The Senior dohc Velos were fancied for a place rather than a win. On form the Porcupine AJS was the favourite machine, having already demonstrated that it had the speed and that its troubles were more niggling than serious. With a team like Graham, Doran and Frend there was every chance that one would win through. Their real danger was "Fearless" Bob Foster on the wide-angle vee-twin Guzzi, a lone dark horse. And so it was, at first. Graham and Frend went into the lead but Foster soon caught them, actually tying with them both for first place on the second lap before pushing off to build up nearly a minute lead by the fifth lap. The Guzzi gearbox failed and Graham looked set for an easy win. Frith, meanwhile, was out, the bevel drive having failed on his third lap while he was lying eighth, but Lyons was soldiering on in his vacant spot.

As all who followed the post-war TT series recall with some sorrow, Graham was robbed of victory when it was almost in his grasp. The magneto armature sheared at Brandish, an almost unheard of failure. Previously Frend tried too hard — twice he tied with Graham — and threw his Porcupine away, and Doran's went sick. Once more the outdated but indomitable Nortons thundered into the lead, Daniell first and Lockett second, but there, hard behind, was Ernie Lyons on the other big Velo which had run with the regularity of a railway train and finished in immaculate mechanical condition.

Of all the races which Bertie watched from the pits, seeing the dohc engines defying the might of the Norton and the AJS camps to win in 1949, both the individual European Championship (Frith) and the Manufacturers Championship 350 cc class, the race that stands most vividly in his memory is the Swiss GP. It started with disappointment for Frith's Velo — too much flooding, perhaps; and before it burst into song after a long push the whole field had gone, the AJS team led by Graham in the lead. Furious with frustration, sweating and puffed out, Frith went completely "fey". He tore after the back markers like a man possessed, cutting through them like a knife, as icy determination gradually replaced the heat of anger.

"I thought it was all over when he got such a bad start," says Bertie, "but in no time at all he came round in sixth place and was gaining on the leaders. It was quite unbelievable. A few laps later he was in front trailing a string of Ajays and weaving from side to side to shake them out of his slipstream."

"I must have been crazy," said Frith. "There I was, a married man of more than 40, with a family, riding like a madman. It was one time when the double camshaft really paid off. I could hang on to third up long climbs right up to 8,500."

On some circuits Frith thinks the single-knocker engine with a wider but lower power band was the best. He found this out the hard way in the Dutch TT. Graham set the pace on the work's 7R but soon Bob Foster had forged through to second place while Frith was unable to get past Doran on a 7R and Lockett on a Norton. The AJS effort soon faded, with Graham retiring and Doran falling off, but Foster was out ahead and going like a bomb. Not on a double-knocker but on a single-knocker —"rocker" engine. Frith caught him but he could not get by, and it was a neck-and-neck struggle until the last corner of the last lap, when Frith outwitted Foster and "pinched" the best line for the dash to the finish. Foster was not best pleased and Frith is not altogether proud of the way he "carved up" his rival. It was the kind of incident which happens all the time in short-circuit racing today — the cut-and-thrust which makes short-circuit racing — but it was trifle unsporting in an era when there was still a bit of "after you" chivalry left.

It's the biggest wonder Foster's Mk. 8 lasted the distance. Bertie recalls a pre-race panic.

"We had the head off for some reason or other and someone turned the engine, unknown to us. When we started it up, it fired a couple of times and then stopped with a nasty noise. The timing was out (that hunting tooth in the bevel drive. The timing marks only line up correctly once in 23 revolutions of the engine as countless Velo owners have found to their cost). The inlet valve had hit the piston and knocked a bit of the top land out, exposing the top ring."

"I had to decide whether to put in a new piston with no chance of running it in, and a good chance of it seizing, or risk the damaged piston. I decided to use the old piston, and it worked."

For the Belgian GP which could clinch his European Championship — he was only one point short of the maximum score but someone else could still force a tie. — Frith, too, chose a rocker engine and beat Foster by 16 seconds after they had eliminated the AJS and Norton opposition.

Two stories concerning Joe Craig from Bertie Goodman's memories:

"After we had won the 1949 Junior TT we had of course to remove the head for measurement but the double-knocker head could not be removed with the engine in the frame. Joe

Study in style (3): Bob Foster, runner-up to Frith in the 1948 and 1949 Junior TTs, was European Champion in 1950

Study in style (4): Les Graham on the last of the line of works racing Velocettes, the 1952 five-speed dohc two-fifty with girder forks and turbo-cooled front brake. He finished fourth in the Lightweight race

Craig walked over and offered us a hacksaw.

"Joe always made sure that nobody looked inside his engines when they were dismantled for measurement, but at Albi where they did not usually bother they suddenly insisted on doing it. Joe had left and the officials passed the Norton head round for inspection. It was most interesting."

What made that season on the Continent wonderful for the spectators but dangerous for Frith and Foster and nerve-racking for Bertie Goodman as the Velocette racing manager was the fact that rivalry was intense between the Mansell-Spring syndicate and the Dick Wilkins camp. Until Frith had got enough points to sew up the individual championship he was in mortal combat with Bob Foster. If both men had been riding in a work's team there would no doubt have been "an arrangement" to share the spoils without risking blowing up each other's machine and risking their necks unduly. But there was no such agreement and Bertie was caught between the two camps.

He could see the danger of the situation but there was nothing he could do about it, for they were not his riders.

Similar rivalry went on in 1950. The Mansell-Spring equipe fielded a bigger team. Reg Armstrong, Ernie Lyons, Frank Fry (who had given Frith a run for his money in the 1949 Ulster) and Charlie Salt, while Wilkins retained Bob Foster and a dashing youngster, Bill Lomas. Bertie did a bit more development on the 350 engines, mostly with cam design (he made an adjustable master cam by slicing a cam in half so that by rotating one half relative to the other, the overlap could be varied); and exhaust pipe length was the subject of experiment too. To allow a non-stop run through the TT, pannier tanks holding 8½ gallons were used. Foster was eliminated from the race by the freak failure of a brake arm and Salt by seized rear-wheel bearings (washed dry by the fuel spraying from the float chamber after the float needle had dropped out). The only bright spot about the TT from a Hall Green standpoint was that Armstrong finished sixth in the Senior on a dohc 500, but things were sorted out later and Foster went on to win the 350 cc European Championship, and the firm again won the 350 cc Manufacturers' Championship.

For 1951 Velocettes once more had a team of their own; Foster, Lomas, Sandford. Foster wanted a 250 and during the winter of 1950, Bertie built a batch of lightweights utilizing the dohc head and an MAC barrel.

Yes, it really was a MAC barrel, but by this time the cooking model of the Velocette range sported an alloy head and an alloy barrel with an iron liner. New lightweight welded frames of semi-featherbed design were used for the work's machines. There was no top tube as such but an adjustable detachable strut ran through a tunnel in the tank to brace the steering head. Early on there was trouble with broken-down tubes but when the fixing bolts for the strut were provided with tapers for positive location the breakages stopped. Working clearance of the bolts in the holes had set up enough vibration to cause the failures. Brake pedals cast in light alloy caused some raised eyebrows, but gave no trouble and were later used on the Thruxton.

Things did not go well in 1951. The 250s had a lot of teething troubles and only one of the three started in the Lightweight TT. Bob Foster rode it but went out on the second lap when the new five-speed gearbox went wrong. His consolation was sixth place in the Junior. Lomas showed early promise by finishing fifth, Sandford retired.

The five-speed gearbox looked quite normal outside but inside an extra pair of gears was actuated by a third selector fork sliding on the upper selector fork spindle and controlled by a third cam groove in the rotary selector plate.

On the Continent, big-ends gave trouble and had to be changed for each meeting, which meant a great deal of work in difficult conditions. Between coming back from the Continent and entering for the Ulster, Bertie modified the big-end by adding two extra rollers, which, as he pointed out, is not an easy as it sounds. Much work on breathing with the aid of a wooden mock-up of the head and a flow-meter eventually got the revs of the 250 up to 9,000 and the speed was pushed up from 100 mph to 115 mph One engine performed unexpectedly well and gave Lomas an easy win in the Hutchinson 100 (the inlet valve was bent when it kissed the piston in practice but was straightened effectively by running the engine at a steady 6,000 rpm for a while on the stand).

Les Graham must have been very surprised when Frith forged his way past to win the Swiss in 1949, but by a twist of fate he, Graham, was to win the same race on a Mk. 8 in 1951. In that year Graham was signed with MV to ride the big four-cylinder and the lightweights, but needed a 350 to make weight. Originally he had intended to ride an Earles-alloy-frame special with a Norton engine, but this plan fell through and Reg Dearden loaned him a Mk. VIII. Dearden had tuned this engine to some purpose... had increased the carburettor size to 1 13/32in from 1 7/32in, fitted a bigger inlet valve, and done secret things to the valve operating gear. It flew. Got past Geoff Duke's Norton at Silverstone, went on to win the Swiss, and picked up places in most of the Continental classics. Altogether, says Dearden, it won £5,000 in prize money and bonuses for Graham that year.

Dearden kept the Velocette in memory of Graham, a great rider, a grand sportsman and his good friend. Money will not make him part with it. Dearden was one of the great "characters" of the racing scene in the 50s. No one can cap his claim of having provided over 1,000 machines for riders in the Island, having helped 11 world champions on their way, and having brought home 229 silver replicas. He had a warm spot for Velocettes, though most people associated him more with Norton.

He thought the narrow Velocette crankcase with the engine sprocket so close (he ran his engine sprockets within 6 thou of the case) was the basic reason for the long success of the design and greatly admired the way the engines were made and machined.

An instinctive tuner rather than an empiric designer,

A privateer's garage! Mike McGeagh (right) uses a packing case as a work bench rebuilding his Mk VIII KTT after a practice blow-up for the 1951 Manx GP. His cartoon (left) is self explanatory! Helping him is Ross Porter and Harry Gill.

Reg would liked to have fitted a Norton top end on a Velocette bottom end, then he would then have the best of both worlds. And though he vouchsafes no details beyond admitting that it is a simple modification to the rocker mechanism which ensures that the inlet valve returns squarely to its seat instead of being tilted slightly by the angularity of the normal rocker, he had a trump card. It was used on the machine ridden by Les Graham.

He explained that a modified valve spring retaining cap was guided by the rocker box so as to restrict fore and after movement of the valve in its guide. It achieved something of the advantage of the bucket tappet design popular with racing car engine designers.

Velocettes' final fling in grand prix racing came in 1952 with Les Graham riding in the Lightweight and Junior, supported by Lomas, Salt and Sandford. The Junior was disappointing for Graham was eliminated when his chain came off at Governors in the early stages and Salt was put out by chain trouble on the third lap — Vic Willoughby, a private runner, who had won the 350 class of the 1946 Ulster GP, had his chain part at Ramsey. There must have been a bad batch of chains that year. Sandford won through to finish ninth. In the Lightweight, Graham on the five-speed dohc Velocette did all that could be expected of him and more than most people expected by coming home fourth in the middle of a phalanx of nine Moto Guzzi riders, led by Fergus Anderson, Enrico Lorenzetti and Sid Lawton. This was a really good show, for the Moto Guzzi was the latest development of a design which had dominated the class, having already won the Lightweight six times, and was backed by the most efficient racing department then extant. The Velocette, though ridden by a star, was very much a shoe-string effort and an adaptation — dare I say a clever codge-up? — of a once discarded dohc design adapted to a basically 350 engine reduced in capacity by the use of a barrel from a 350 roadster and flywheels from an obsolete 250 roadster (the MOV). It had one noticeable improvement since appearing the previous year. The front brake was a turbo-ventilated 2ls design intended for the Velocette four.

Velocette four... What four?

Well, that was one machine which never was. Or was no more than an unfinished 125 cc single-cylinder test rig and a very advanced design of front brake in which internal vanes were used to draw air over the finned brake drum, the whole unit looking and working rather like a child's humming top.

Bertie Goodman once stated that the whole thing started as a leg-pull in a hotel lounge at Berne in 1949. Meeting Joe Craig, he asked pleasantly how the Norton four was getting

Freddie Frith: another shot of his 1949 Junior-winning ride

on. Now it had been rumoured that Nortons were planning a four-cylinder racer (in fact the BRM organisation at Bourne were building one for them and the McCandless designed featherbed frame was intended for it), but Joe Craig was not to be drawn and Bertie thought he would give him something to worry about. "Wait till you see ours — it will give Bracebridge Street something to think about," he said as a parting shot.

Bertie swears that until that moment there was no thought of Velocettes producing a four but, as often happens, the jest started a train of thought. They were at the time, top of the 350 tree, but it was still the Junior class, prestige wise. Senior laurels had always eluded them. Could they perhaps rehash the 500 cc Roarer twin? Back home he talked it over with his father. They sketched out a revised version with liquid cooling (the original pre-war design had been for water cooling but was changed to air cooling in the interests of simplicity and compactness), but before they had got far they realized that potentially a four would always beat a twin.

A four it would have to be. Something of the old racing design fever which had been stifled for many years during the war and diverted into the pedestrian channel of the LE must have come back to Percy Goodman. Bertie, with his experience of the current GP scene, set out the requirements and his father sketched out a design to meet them. The target was the 1953 TT and the first task was to build a single-cylinder 125 unit from which practical data could be obtained.

"We reckoned," says Bertie, "that we could get 85 bhp at 12,000 rpm Having done the design, my father got an outside draughtsman to do the drawings. There is always a big problem when you go racing. You tend to put your best brains on designing the racers and the production jobs get neglected." (Both of us thought immediately of one classic instance of production machines stagnating while racing took place.)

The Velocette four as committed to paper was a square bore and stroke (54mm) double ohc design in what has come to be regarded as the classic type. Apart from having roller bearings for the mains and big ends, you can picture it as a slice off a miniature XK Jaguar engine, even to the inverted piston tappets enclosing the valve springs. The transmission was interesting, for Percy again insisted on shaft drive as used on the LE. But this time the crankshaft was across the frame and the in-line gearbox and clutch was driven by a bevel on the clutch shaft meshing with a bevel combined with a spur gear meshing with the crankshaft gear.

An all-welded frame with single top tube, well-gusset-

Old KTTs never die... they are used in the Manx GP. Manxman Michael McGeagh takes his 13-year-old KTT into Parliament Square in the 1952 MGP

Bertie Goodman relives his past at a Vintage Club's Mallory meeting

ted steering head and widely-splayed duplex cradle was planned and in the subsequent 200 cc ohv Valiant one can see traces of the same thought. If you picture a Valiant frame with front down tubes and cradle splayed to the width of a Granville Bradshaw vintage ABC you will not be far off the mark. But before the experimental single-cylinder unit was completed, only a week before, Percy Goodman, who had been failing in health for some time, died. The board of directors reviewed the situation and decided that there was no future in racing as far as Velocettes were concerned. It had become, as other firms were to find out, a very expensive business and in their particular case, the publicity accruing from racing was not likely to have much effect on the sales of their current products, the commuter LE and the touring MAC. Bertie thought then, that the estimated cost was a little high and much of it might have been retrieved by way of petrol and oil bonuses, but he was the only one really sorry to see the racing side dropped. So what might have been a "world beater" (this is a contemporary expression. It used not to be considered in good taste to talk of victory until you had received the chequered flag, and then only half apologetically) was to be still born.

"Was the Velocette four to be the basis of the Manx Lottery world-beater project?" I asked. "Definitely no," said Bertie.

After the death of his father Bertie became Sales Director, a rather unconventional sales director given to rushing about the country on a MSS whenever he could find the excuse, and jumping at every chance to do a bit of high-speed road testing. When the demand began to build up for bigger banger scrambles machines — particularly from America — it was just his cup of tea. He soon set about putting some steam into the touring MSS. Upping the compression ratio and fitting a cam based on the Mk. 8 profile transformed the MSS into a potent piece of machinery and Bertie enjoyed himself no end bashing around scrambles courses and bombing around MIRA. And this is how the humble touring MSS was eventually transformed into a production racer and world-record breaker.

The scrambler was never really much of a success. The engine earned a good reputation but the frame was not developed to match.

"I made the mistake," said Bertie frankly, "of using a standard frame and of not employing star riders who could evaluate the machine." In any event the writing was soon on the wall for all big banger scramblers.

But this episode had given him a renewed taste for the competition world and he next became interested in the Thruxton nine-hour race. Interested because he could himself "have a bash".

He learned a lot about Velocettes the hard way at Thruxton, including the fact that the oil-bath primary chaincase did not retain oil under racing conditions. What happened (before the primary chain ran dry and broke) was that the gearbox plates flexed, distorted the cork seal in the case behind the clutch, and let the oil out. Thereafter a double-thickness drive-side gearbox-engine plate was used.

Happy hours bombing round MIRA flat out kept his weight down and developed the Venom into an extraordinarily reliable production racer. When, therefore, a French syndicate of sportsmen headed by the Monneret family suggested that they should have a go at long-distance world's records at Montlhery Bertie was keen, and confident that the Venom could stand the hammering. Equally confident that for this kind of job a well-developed roadster was a better bet than a racer bred for a short life and a gay one. He had some justification for this belief because back in the 40s he had planned a pre-Earls Court publicity attack on some long-distance records with Freddie Frith riding a Mk. 8 at Montlhery and it had been a dismal failure. The engine had produced enough power in an endurance run on dope on the brake, but when they got to Montlhery they found the difference between theory and practice. For one thing, the centrifugal force created by riding round a banked track soaked up power. For another they could not get the engine hot enough. It was cold weather and with the big alloy finning and the dope fuel it ran too cool. Would have been better with an iron engine. It was one record attempt which did not get into the weeklies!.

It took five months' development, and Bertie covered 2,000 miles at MIRA. before the Venom, so nearly standard as to be unbelievable, was ready. It was undoubtedly the fact that it was so standard, and that every part was well tried, which resulted in such great success. One other factor came into it, and the dominant lesson of his personal experience as a rider, was the importance of cutting down the rider's wind resistance.

"I suppose I had a thing about wind resistance - I was probably the first rider to use a one-piece suit — it was a German paratrooper suit actually. Testing at MIRA taught me that the difference between tight leathers and baggy ones could be as much as 6 mph. Wearing gloves with cuffs or turning the toes out could make a noticeable difference. Reducing the thickness of the seat padding by 1in, which would only be ¼in compressed, could make a difference of half a second a lap.

What four!

VELOCETTES' connection with not one but two Grand Prix racing four-cylinder projects was a sad and sorry story.

The sad part is that the first one proposed back in 1951 was Percy Goodman's last dream of putting Britain back in the World Championship, but he was taken ill and died before a prototype was built or the 125 cc single-cylinder test piece completed. With his death it was not surprising that the rest of the Board decided against going on with such an ambitious and expensive project. News of this four was kept pretty quiet for obvious reasons and few at Velocettes knew anything about it for the good reason that the detailed drawings based on Mr. Percy's outline were not done at Hall Green and not even by a Velocette employee. Percy had engaged George H. Jones to do the work at his Shrewsbury home because, as he told Jones, "If I give this work to our works drawing office there would be little else done and we have other urgent things to do."

Being liquid-cooled (glycol), the unit was only 14½ inches wide. The widely splayed frame tubes enclosed a small pressurised radiator within the frame, the in-line gearbox provided five or six speeds and final drive was by shaft like the Roarer. Electronic ignition and fuel injection was planned and the design target was 70-75 bhp at 12,000 rpm to give 175 mph. Weight, with lavish use of magnesium alloy and titanium, was forecast as not more than 325 lbs.

Jones, one of the industry's most versatile and talented designers never sought publicity and his work was never known outside a small circle but he gave us the 'lighthouse' vertical camshaft OK Supreme 250 cc of 1930 which was capable of 8,000 rpm, an almost unbelievable performance in those days, the special vertical carburettor cylinder head for a JAP which enable OKs to win the Lightweight TT in 1928 (an idea taken up by BMW cars in 1938 and used with great success); the JMB three-wheeler of 1933; the Villiers 98 cc autocycle unit and several other moped designs. Recalling his work on the Velocette four he wrote "It was a very pleasant assignment with frequent visits to Hall Green and an honour to work with such a fine designer and gentleman. I have no doubt that it (the four) would have proved a worthy competitor to the continental machines." The Japanese were not then in sight.

No word of the abandoned four leaked out until the sorry part of the four story began, with a grandiose plan for an Isle of Man lottery which was to raise money for local hospitals and provide a fund to sponsor a British grand prix motorcycle to take on the world. The lottery — copying the Irish Sweepstake — forecast it would be able to donate £25,000 to the fighting fund by 1966 (quite a lot of money then). This money would be administered by a committee made up from the lottery organisers, the motorcycle industry and the ACU. As one may imagine the committee had difficult questions to answer. What bike, what manufacturer, what classes? were the main problem. There was a tremendous amount of talk but little action but a hopeful sign was that the committee would be limited to three, someone from the lottery organisation, someone from the Industries Association (Hugh Palin) and someone from the manufacturers (Bertie Goodman). The weekly magazine *Motor Cycling* invited suggestions for a 'world-beater' design from designers and there was an avalanche of ideas, from the strictly practical from Phil Irving (a three-cylinder slice out of the BRM V8 Grand Prix car engine) to the downright outrageous from Philip Vincent (a V8 liquid-cooled 250 dohc four-valve-per-cylinder *two-stroke*). Vincent forecast 100 bhp at 20,000 rpm but at least he added he didn't think there

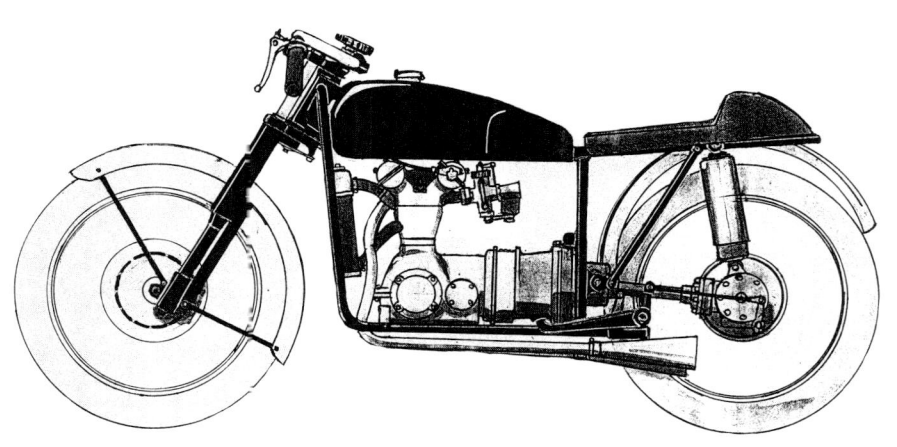

Dave Master's impression of the proposed Velocette four

was a chance in a thousand that such an unorthodox design would be considered, let alone adopted. Several manufacturers' designers came up with ideas which reflected their firms' interests and there were some interesting outsider designs but Bertie Goodman really trumped all their aces by announcing that he had a design in his desk drawer that might fit the bill. He had, too, for it was his father's design and he argued that to use it would save a lot of money — £24,000, he estimated — and the £20,000 offered by the lottery was not enough to fund a new design.

When after months of rumours the first lottery run on the Senior Manx GP in 1965 paid out £79,000 in prize money (a Douglas woman won the first prize of £25,000), £7,000 was allocated to the world beater fund. It was going to be a long time before the fund would reach the £25,000 which Palin of the Industries Association considered was necessary to build a bike, but later in the year it was announced that a secret backer had offered practical and financial help, and all parties had agreed to press on with the plan. There was no hard news until early next year (though there were rumours of discontent in the Island about the running of the lottery) when the announcement came that Sir Alfred Owen, the industrialist who had bought the BRM race car firm, was prepared to let BRM join forces with Velocette to build a bike to put Britain back on the racing map and Bertie Goodman would be in charge. BRM, it emerged, was to be responsible for the engine and Velocette for the rest.

Enthusiasts now felt more optimistic; Sir Alfred was known to be very patriotic (hence his taking over BRM) and a man to get things done. We figured he must have been the secret backer. We all had confidence in Bertie Goodman's position, in charge. He at least knew what a racing bike should look like. We would have liked to know what form the bike would take though we appreciated the need for a certain amount of secrecy. Was it to be a four-stroke or a two-stroke? A four-stroke surely, because BRM had no experience of two-strokes. We never did find out!

It did not really matter because the lottery was in trouble. Allegations and accusations were flying around in the Island and its Lieutenant-Governor appointed a special committee to " ensure that there is no cause for public disquiet over the operation of the lotteries." This was widely interpreted to mean there *was* cause for public disquiet and ticket sales for the next lottery were right down. After considering the report of the first inquiry the Governor ordered another one to consider changes in the law over lotteries. This would take time to go through the Manx Parliament and when the results of the second lottery were seen to be down, and with it the prize money, the lottery committee resigned after a row with the Government, which announced there would be no more lotteries for some time to come. Apart from some claims going through the Island courts by an agent for his commission on the sale of tickets, that seemed to be the end of the lottery scheme with its high hopes.

It is unlikely that any real work was done on the project, especially as to this day no one, not even Bertie Goodman, seemed to know what form the engine was to have taken. Bertie, quizzed many years later, thought it was to have been a two-stroke 250 cc four. A BRM engineer at the time thought it was to be a 500 cc four-stroke four. That makes more sense for in the last days of Norton there was a joint project (with BRM) to build such an engine. That too got no further than a single-cylinder test piece.

Ralph Seymour, of Thame

RALPH SEYMOUR* (R.F. Seymour Motor Cycles Ltd., Thame, Oxon.) was known by name to Velocette enthusiasts the world over yet few have met this quietly spoken, modest man who did so much to keep the Velocette flag flying on the race track and road. I have known him for more years than I like to remember (it must have been in 1948 when he first gave me a warm welcome).

He was then running the motorcycle side of a large car and motorcycle business and I was a greenhorn motorcycle accessory rep. getting used to being rebuffed by buyers only interested in money. Ralph was not like that. He could not give me much business but we talked bikes, vintage and post vintage mostly. Grindlay Peerless, Rudge and Velocette and I set off on my lonely travels on a solo,

* *These words were written a few weeks before Ralph Seymour's death, at 78, in February, 1994. C.E.A.*

Ralph Seymour with his original Mk VIII KTT; Manx Grand Prix, 1948

my spirits revived. I have never forgotten that meeting.

Ralph Seymour started off as an enthusiast and remained so through the ups and downs which bedevilled the motorcycle business. Riding a vintage Rudge Ulster, he was apprenticed to the hard stuff at the wonderful Clubmans Day events at Brooklands in the mid thirties.

Promoted by *The Motor Cycle* (the Blue 'Un of fond memory) and organised by the Brooklands Motorcycle Club, this Day gave ordinary clubmen a chance to have a go on ordinary road machines. The rule demanding the official Brooklands silencer was waived provided you had a standard road going silencer though not all of these were, I recall, standard and a lot of the lads fitted Brooklands cans because they looked more the part. There were lots of short races and speed trials over a

Ralph with his tele-fork Mk VIII KTT at the Manx Grand Prix, 1952

kilo. Entry fees were cut price and there a few serious races for BMCRC members (the professionals) to show how it should be done.

Literally hundreds of clubmen who rode at these meetings went on to be serious racers. Ralph Seymour made friends with Arthur Wheeler and David Whitworth... many friendships were made and reinforced at race meetings for years afterwards.

He remembered with a wry smile that in 1936 his home-tuned Rudge Ulster clocked 80 mph over the kilo and next year, with a pukka ex-work's Graham Walker one, he managed only 78 mph.

His 'Velocette saga' really began when he bought a Mk. 8 in 1939. It cost him £97 (trade price), part of the money coming from the sale of the Rudge to Harold Hartley who rode it in the TT. The Mk. 8 now wears telescopic forks, a works front brake and an ex-works 7-gallon tank, of which more later. There are three other Mk. 8s but they do not really count. He became a regular rider in the Junior Manx after the war on this machine (finishing 38th in 1946, 21st in 1947, 15th in 1948, with one retirement in 1949). On a Norton in the Senior Manx he had two retirements but was 12th in 1948 and 1949. He then switched to the Junior TT, his ambition being to earn a replica. He was 45th in 1951 and retired in the Senior on a Junior Velo but made a special effort for 1952.

"I asked Bertie Goodman if I could have a pair of tele forks they were using on their new works bikes [Bertie had already let him have a special 2 ls front brake and a 7-gallon tank]; he said I could have a pair but a Mk 8 would not steer with them. I went home and made my own yokes. When Bertie saw my bike with them on he said someone at the factory must have given away the information. They had not. What I had done was to do a scale drawing from a photo of a work's bike and work out the angle."

"Was the steering better than with girders?" I asked. "No, not really, it understeered but the braking was better."

He finished in 38th place in the 1952 Junior, gaining his second TT replica to go with his three Manx ones; that was his last ride.

The firm he worked for were losing interest in motorcycles as the 60's approached and he decided to start up on his own. Premises were a problem with his limited finances, and the fact that dealerships were tied up in most areas but here his love of Velocettes paid off. His friendship with Bertie Goodman paid off too and he got the Velocette agency. Triumphs turned him down but Neale

"Ralph's finest hour". Fred Walton on Ralph Seymour's Thruxton Metisse at the Bungalow, on his way to 19th place and a silver replica in the 1973 Senior TT. The white scar on the fairing of the Thruxton Metisse, near the megaphone, is a result of 'enthusiastic' cornering at Kirk Michael.

Shilton, the area rep. and a great motorcycle enthusiast, got the decision overturned. A customer had a KSS that had been overhauled by a prominent London Velocette agent but had a baffling oiling problem. Ralph's workshop foreman would not touch, it but to help the owner out Ralph agreed to do it at home in his spare time. He solved the problem (the oil pump had been fitted incorrectly). Told that Ralph was looking for premises, the owner came up with the advance information that the owners of Hawthorne Works at Thame were moving out. The price was reasonable and Seymour's are still there.

It's a rambling Victorian building and the showroom where a modest range of Suzukis are available for local customers are not pretentious but upstairs is the workshop and spares department, run by wife Eileen and daughter Liz, whilst downstairs are the machine tools which enabled Ralph to make the parts for obsolete Velos which keep their wheels turning the world over.

"At the liquidation sale at Hall Green other people bought machines and parts but with my limited resources I decided to buy machine tools so I could go on making things when the supplies run out. But I also bought the valve spring tester Harold Willis used for £5 and the clock up there on the workshop wall came out of his aeroplane."

No longer having the time to race and having now too many responsibilities Ralph began to support (he corrected me when I used the word sponsor) riders on his Mk 8s and later his Venom and KTT Metisse specials. Call them what you like, when the last generations of Goodmans at the works, Bertie and Peter, saw the first of the line they agreed that he

Fred Walton at Brandish on 'Katie', the Mk. VIII KT- engined Metisse

could register it as a Velocette. With the well tried Venom engine in various stages of tune in a Rickman lightweight duplex frame, all welded with no heavy lugs it was of course the bike that serious Velo men wanted. It was the bike that Velocettes might eventually have made if their funds had not run out. The project began when, trying to make his old Mk. 8 competitive, he fitted the engine and a hush-hush five-speed gearbox (the Goodmans had always given special bits to help riders 'adopted' as members of the big family of Velocette enthusiasts). This machine evolved into 'Katie', Ralph's pride and joy.

To fit the readily available and tuneable Thruxton engine was obviously the next step and in no time he was being pressed to start a production line of racers. Demand for a road version followed. In all sixteen racers and half a dozen roadsters were built, all to order and as soon and when they could be fitted into the workshop workload Much of the work was done in Ralph's own time after shop hours and, because he loved the work, the real commercial cost was never, I think, included in the pricing. But as anyone who has built a bike will agree, no price can be put on the satisfaction of seeing your creation in action. You can imagine the pride with which he saw his Velocette Metisse machines come home in good places and when he recalled how Fred Walton finished 19th in the 1973 Senior TT he seemed embarrassed by the feeling of pride which welled up and toned it down with excuses: "We were lucky, a lot of riders retired" and "Fred really got into the groove."

The roll of honour of riders who were 'supported', not sponsored, on either Mk 8s or Venom Metisse models is lengthy but apart from the afore mentioned Fred Walton the following were called to mind: Gerry Edwards, Ted Hunt, Alan Holmes, Ron Bryant, Norman Price, Vernon Wallis, Keith Edwards, Dave Skelly, Peter Kermode, Danny Shimmin and our publisher, Bill Snelling.

The Metisse days are long gone (although a roadster with full-road going equipment was being worked on in 1993 for a customer in Japan). He found great pleasure in his more recent involvement with Bernard Guerin, a French 'garagiste' member of the Vintage Club who races a Seymour-prepared Mk. 8 in that club's races when ever he can. Bernard's greatest success in the Isle of Man was winning the Senior Classic on the Billown Circuit in 1993.

The friendship between Guerin and Seymour began when a letter arrived from France asking in French if Seymour could supply the parts marked with arrows (to overcome the language problem) on an exploded drawing (it was of an MAC though the parts were for a Mk. 8). Ralph got a schoolboy to translate the letter, the parts were supplied, and that was the start of a wonderful friendship, a friendship reinforced by an all-night rebuild of the Mk. 8 before a race meeting.

Seems, Bernard had rebuilt his engine before coming over to race and they took it to an airfield to give it a run and set it up. Near the end of the session it tightened up. An incorrect spacer had seized on the timing side and defied removal. It had to be cut through with a flexy grinder which ruined the shaft and bevel pinion. The engine had to be stripped and built up with new parts. They worked all night and finished just in time for breakfast.

Such tests forge real friendships.

The latest Velo victory in the Island, and the first French victory. Bernard Guerin splashes round Castletown Bridge Corner en route to a win in the Classic TT Vintage race on the Southern 100 course, in June 1993, on Eileen Seymour's Mk. VIII KTT. Bernard's favourite saying was: "Always tomorrow". But not after that ride!

DECEMBER 1952

119.87 m.p.h. on a 21 in. VELOCETTE

...a new American 21-inch Record!

BONNEVILLE SPEED TRIALS
August 1952

MAC VELO only $599.50
Inc. fed. tax.
F.O.B. Los Angeles, Baltimore, Houston

LLOYD BULMER "flat out" on his record-breaking 21-inch VELOCETTE

On August, 29, 1952 Lloyd Bulmer flashed through the Bonneville Speed Trials official timing traps on his 21-inch VELOCETTE at an average speed (for both directions) of 119.87 mph—a new official A.M.A. 21-inch record. This is further proof of the superiority of Velocette motorcycles.

ASK YOUR DEALER TO SHOW YOU THE NEW MAC...

The latest MAC 21-inch O.H.V. will gladden the hearts of all true motor-cyclists. Fine quality and superb design are characteristic as ever of this famous machine.

Backed by over a quarter of a century of successful racing experience, Velocette engineers are able, in the light of lessons learned, to build into their production models those subtle differences in design that distinguish the Velocette as a real thoroughbred.

"A Thoroughbred from Start to Finish"

New... HANDY COMBINATION KNIFE & TOOL KIT

WRENCH opens to cutlery steel KNIFE, AWL PUNCH, OPENER, EDGE FILE, PHILIPS SCREWDRIVER, FLAT FILE, SCREW DRIVER

FREE! POCKET MOTORCYCLE KIT $12.50 Value

One of these new sturdy Pocket Motorcycle Kits will be given free to every buyer of a NEW MAC VELOCETTE. Kit is hand-forged of finest imported tool steel, finished with a rust-resistant, gleaming NICKEL plate. Excellent for emergency repairs on the road, trail, etc. All you need for tune-up work. Tool kit comes packed in handy genuine leather carrying case. C.O.D. MAIL ORDERS accepted at regular nationally advertised price of $12.50 per Kit.

MAC SPEED KIT $75.00

DEALERS! Velocette means profits for you. Write today to BRANCH MOTORCYCLE SALES for choice territory.

DEALERS WANTED

BRANCH MOTORCYCLE SALES   C-12
2019 W. Pico Blvd., Los Angeles 6, Calif.
Please send me descriptive literature on the:
☐ VELOCETTE   ☐ MOTO GUZZI

NAME _____ (Please Print) _____ AGE

ADDRESS _____

CITY _____ STATE _____

BRANCH MOTORCYCLE SALES
2019 W. Pico Blvd.   Dept. C-12   Los Angeles 6, California

Post-war roadsters
"Bread and butter" activities at Hall Green

Preoccupation with Velocette fortunes on the track has caused some neglect of this side of the story, covering the "bread and butter" activities at Hall Green, which made the racing programme possible. I will now remedy this omission lest it be thought that it was all play and no work. In fact the racing side of the company's operations was so small that it is remarkable that so much success was achieved in the immediate post-war years.

Post-war production began in 1946 with a batch of 200 GTP two-stroke models, all magneto equipped, all of which were exported, presumably they were assembled mostly from pre-war parts. These were the last of the two-stroke range that started in 1914. Thereafter the pre-war range of MOV, MAC, KSS and MSS models was reintroduced and manufactured in small batches. In October 1947 the KSS was discontinued. Some work had previously been done on the KSS to raise the performance and a larger inlet valve fitted. The result of this work was probably seen in the models which were run in the Clubman's TT. Ronnie Hazlehurst won the 1948 Junior Clubmans on a Mk. II KSS, but in the ordinary way the post-war "cammies" did not seem to be any different from the pre-war ones. Production of the KTT was resumed, as has been mentioned earlier.

Models MOV, MAC and MSS and the KTT were continued in 1948, Dowty tele forks being used on the roadsters. Models MOV and MSS were discontinued in September and the LE 150 was exhibited at Earls Court in November.

In 1949 only the LE and the MAC were produced but in 1950 there were some changes in specification. The LE was increased to 192 cc in September and major modifications were made to the MAC which had been unchanged apart from small details since its inception in 1934. Telescopic forks of conventional design and Velocette manufacture replaced the Dowty air-sprung forks. They appeared on export models only at first, and from July a new cylinder head in light-alloy with fully enclosed valve mechanism and an alloy jacketed cylinder went into production. The object of the engine modifications was not to raise the performance so much as to modernize the unit by utilizing contemporary alloy casting techniques, techniques in which

All-alloy MSS engine, a design that stretched from the 23 bhp MSS, through Venom and Viper models destined finally to become the basis of the 38 bhp Thruxton. Illustrated is what appears to be an experimental pre-production unit; none of the production Udall-designed engine units have this style of rocker drain tubes

Velocettes had pioneered the way with the Mark 2 KSS and the later KTT models. It was a tidying-up operation which transformed the rather spidery and "bitty" looking iron-head MAC into a more impressive, all-enclosed engine with that "smooth" look which was becoming the fashion. This was achieved by enclosing the valve mechanism in what I can best describe as an aluminium casserole. I use the description deliberately because any attempt to develop this engine beyond the maker's intentions was likely to lead to "cooking" of the valve gear. The cylinder head formed the lower half of the "casserole", the sides being brought up nearly vertical all the way round and finished level.

The top of the "casserole" was an alloy lid and no more. It was held down by a series of small bolts, was quite smooth, and devoid of any finning and, being polished, added little to cooling. The rockers operated in split-alloy housings which were bolted down to pillars rising from the floor of the "casserole". Three studs alone held the rocker

Post-war Velocettes at an Australian show. All models, including a KSS, rear right, are fitted with Dowty pneumatic forks

housings down and it was all too easy to strip the threads or shear them off. It was a clean, oil-tight design, ideal for its intended function in a well-mannered, semi-sporting roadster, but not suitable for further development. The "Map of Africa" timing cover was also changed to the smoother outline version that graced all pushrod models until the end of production.

If I may digress a little, it is to mention that dealers and enthusiasts continually asked, "Why did not Velocettes make a modernised MOV 250 based on the alloy MAC?" On the face of it, of course, doing that looked easy. The original MAC engine had been produced by lengthening the stroke of the MOV, and the process could be reversed as easily. By the time the clamour for a 250 arose, the MAC had been provided with a pivoted-fork frame heavy enough to take a 500 motor, so a new frame would have been necessary, but after the Valiant 200 cc twin had been produced quite a few armchair designers, myself included, figured how nice it would be to have a MOV motor in this lightweight "featherbed" type frames.

I put these "brainwaves" to Bertie Goodman and he explained that the limitations of the MAC had been taken into consideration when the need for a 250 arose.

"The alloy head of the MAC was not all that good, we couldn't use the old one because it was too old fashioned, and the difficulty in making a 250 is that customers are only prepared to pay '250' price. We decided to make the Valiant using the LE unit as a base because it was more up to date

Two models for America only: the 348 cc Endurance MAC cross-country model (left) and the MSS Endurance

Adjustment system for pillion load employed on all road-going ohv Velocettes from the early 50s. It is a development of the Phil Irving designed system of pre-war years

and gave us a specification that was better than anything else in its class... unit construction, shaft drive, and the twin sound. If we had been able to make it at a price within £10-£15 of other 200 cc machines I think it would have sold well."

The spring-frame MAC was introduced at the Earls Court Show in 1952 although the rigid version remained in production in 1954, being preferred by American customers. The spring frame was based geometrically on the Mark 8 KTTs which it closely followed ahead of the swinging-fork pivot, and the characteristic taper tube fork clamped to a cross shaft operating in plain bushes was virtually identical to the racer's layout. To retain the advantages of the patented adjustable springing (the 1938 P.E. Irving and Veloce Ltd. patent) and to provide a low riding position either by saddle or twin seat, the seat stays were swept down from the seat lug as on a rigid frame and then curved back to the lower cradle tubes. The pressing provided the curved slots for the adjustable spring mounting were welded to the loop stays. This frame, used on all singles, including the Thruxton, and can truly be said to have stood the test of time.

It is functional and distinctive and provides the steering qualities of the KTT but it is not fashionable in the modern idiom. Other manufacturers, when jumping on the swinging fork band waggon, copied the horizontal seat stay and diagonal strut of the McCandless design used on scramblers and the "Featherbed" Norton, yet *this* frame was a direct copy of the 1936 work's racing Velocette frame. Velocettes did not consider this frame suitable for road use because of the high riding position and lack of easy adjustment for pillion load (the Dowty oleo legs were adjustable for load because the air pressure could be varied but this was not an operation to be trusted to a ham-fisted owner, and in any case the units were too expensive for a roadster). Instead they developed the Irving system of moving the abutment of the spring units, a method which was simple, visible and virtually foolproof and in my opinion infinitely better than the use of adjustable spring units. Far from being "old fashioned", the Velocette frame was ahead of more fashionable designs in this respect and was yet another example of the firm's refusal to compromise their engineering convictions for the sake of fashion whimsies.

The next milestone in Velocette history and one which was to have far more significance than at first appeared was the redesign of the 500 cc MSS in 1953, resulting in the pleasant surprise at the November Show of a 500 cc engine in the pivoted-fork frame of the MAC.

In designing the new engine, Charles Udall could draw on a vast fund of experience of the "M" engine dating back to the early 30s, when he had assisted Eugene Goodman in the design and development of the MOV two-fifty, but then enlarged it to the 350 MAC and had finally designed an enlarged version, the 81x96mm MSS. The new engine could therefore be expected to embody all that was best in the old designs plus the advantages of modern materials and technology.

The crankcase and timing gear remained virtually unchanged. The timing gear, particularly, with its train of fine-pitch helical gears leading to a cam wheel placed high and above the cylinder base flange, had earned an enviable reputation for silence and longevity. The taper roller bearings supporting the flywheel assembly in the crankcase were unconventional but had been used with success in the later post-war long-stroke MSS engines. The use of taper rollers with a calculated pre-load was an unexpected application in a motorcycle engine though normal practice in heavily stressed car back axle differential units. The reasoning for their use in the Velocette engine was that, size for size, they had a higher load-carrying capacity (if for no other reason, than that the rollers be longer, that they were more able to withstand out-of-line deflections inevitable with a flywheel assembly, better able to cope with the end thrust generated by the helical timing pinion; and with the final bonus that they could be given a pre-load which would eliminate play in the assembly over long periods.) All reasons which, expense apart, one might have thought would one have commended them to all other engine designers.

However, the narrow chain line traditional in four-stroke Velocettes may have caused Udall to look that little bit harder into the question of main bearings. Another designer not so limited in space would have added an extra bearing to give sufficient load-carrying capacity and would then have missed out on the other advantages of taper rollers. The most obvious change in the new engine was the abandonment of the traditional long stroke in favour of a square 86x86 ratio. One reason for this was obvious - it made the new unit much shorter, so short that with a squeeze it would fit into the existing MAC spring frame. (A squeeze only in the sense that the cylinder head-barrel through bolts have to be removed before the head can be taken off with the engine *in situ* in the frame.)

The other reason for the square ratio was debatable. Square engines are now obligatory for high-efficiency and the tendency is to over-square. It was not so in the early 1950s and the MSS had never been envisaged as a high-performance or racing machine. At the time of the announcement of the engine Udall confessed quite openly that the stroke had been reduced to facilitate installation in the MAC frame. Current gossip on the Birmingham "grape vine" was that Velocettes had shortened the stroke of their old engine to make it fit the MAC frame and had been quite startled at the way it motored. When a while later, however, I quizzed Udall about the stroke change he denied that it had been done out of expediency. The change was, he said, dictated by his desire to make an engine in accordance with the then latest design technology. Whatever the reason, and

An American dealer, possibly Don Hamilton, with an Endurance model

I believe each of the considerations mentioned came into it, the end product succeeded beyond expectations. I cannot believe that anyone at Velocettes, Charles Udall included, could have envisaged the power that would eventually be wrung from this roadster design and the world fame it would bring to Hall Green.

One factor which formed a sound basis for future development was the detail improvement of the flywheel assembly. For the first time the crankpin was a press fit in the flywheels, a slow taper being employed so that the pin could be entered in the wheels and then pressed up to the shoulder of the big-end track. This pressed-in pin (a feature of the earlier two-strokes) greatly strengthened the assembly, for the bosses no longer had to be counter-bored to make room for nuts. There was nearly twice as much pin supported by the wheels, and it was larger in diameter anyway. Careful thought had gone into the big-end, too. Long rollers, 3/16 x 9/16in thick needle rollers, you might say, utilized almost all the bearing surface of the crankpin, the cage being set into grooves in the flywheel cheeks so as not to waste any of the big-end bearing surface. Finally the oil was led out of a hole on the "inside" of the crankpin, not the "outside," so that the oil would be centrifuged through the bearing and not out of it. The flywheels were, of course, steel stampings, not cast iron. If the crankcase assembly be the heart of an engine, this one had the heart of a lion. What of the lungs?

The head bore some relation to the alloy MACs in that

In the USA: a trio of Velo scramblers, riders unknown, head the field

it was machined flat above the top fin, leaving pockets for the valve springs. The pockets were now shallow troughs for hairpin valve springs were employed, hairpins based on KTT experience, to obtain the required strength with low overall height, and the valves were allowed to rotate, another lesson from the KTT. Where on the MAC the rockers were bolted to the head, suffering from its high temperature and with the holding down bolts having to take the strain of operating pressure on the new design, the rocker box cover carried the rockers in its roof in the familiar split-alloy bearings well away from the heat and spring forces being carried by the widely disposed rocker box fixing bolts. To some extent the principles involved were reminiscent of the post-war dohc racing machines where the cam box was bolted down to a flat surface machined on the head. As announced in 1953, the engine turned out 23 bhp with air filter on a compression of 6.8. Softer cams than on the iron engine were employed in the interests of torque, petrol economy and silence though the

Veloce constructed this one-off trials machine, based on an Endurance frame. It was ridden by John Hartle

power output was restored using cam followers with a longer pad of greater radius which increased the effective opening period. Now with the overlap increased from 38

The unmistakable exhaust note

THAT silencer. Since the early vintage days you have been able to pick the exhaust note of a four-stroke Velocette out from the rest by its dull, though not unmusical, thump, thump, thump merging into a muted roar. Other manufacturers altered their silencers nearly every year in a futile attempt to be "with it", often with little concern for the public or their exhaust valves. Not Velocettes. They evolved a design of silencer way back in about 1926 and stuck to its principles ever after. The early vintage silencer was an unlovely thing the shape of an army water bottle with a tiny, sawn off at an angle, stub of a tail pipe. Internally it worked on the same principle as a Brooklands silencer in that the inlet tube was at a much lower level then the outlet to prevent the gas going straight through. Then, as later, the baffle system incorporated tubes with slots punched at an angle to further confuse the gases. Long ago enthusiasts removed these baffle tubes only to find that the noise went up and the performance went down. The characteristic fishtail shape began in the early 30s, but the silencer still worked on the same principle, and if you "tuned" it you tended to lose performance. Fitting pattern or custom silencers always upset the performance and could ruin the exhaust valve and the fuel consumption. Eventually later generations of Velocette enthusiasts got the message handed down from the past.

Other enthusiasts tried the latest fashions in silencers, like women trying hats, but not the dedicated Velocette enthusiast. When his silencer disintegrated he bought a genuine maker's one. He might have winced at the price compared with some flash-in-the-pan silencer but he knew no other would give the right note and the right performance.

"The odd thing is," said Bertie Goodman, "that although over the years, we tried many other silencer ideas we never came across anything better from a performance, with reasonable silence, point of view than the old vintage-type silencer. Not until much later did a silencer equal it, and that was the megaphone device Floyd Clymer fitted to his Velocette Indian. Even that had Veloce-made baffles fitted in its megaphone."

The Velo silencer was a back-pressure system, unlike the absorbtion type fitted to most machines. A loose slotted open ended baffle fitted snugly inside another fixed baffle in the silencer which had a solid end cap. The slots in both baffles were angled in opposing directions so that the exhaust gases were made to do some contortions before emitting the characteristic Veloce "woof".

degrees to 100 degrees, the compression ratio upped to nearly 9 to 1, with bigger ports and valves, this engine turns out nearly twice the power and, as we and the world learned, could keep it up for 24 hours without showing any strain.

In the final assessment, the redesigned MSS developed through Venom and Thruxton stages and partnered by its little brother, the Viper, goes down in history as a worthy successor to the immortal Percy Goodman "cammy" and the all-conquering Mark 8 KTT.

It would be opportune to deal with the Venom and Viper at this point so that continuity is not lost by excursions into other Velocette diversions.

Experience in uprating the performance of the MSS for cross-country work and marketing it in scrambler form showed that simple tuning, raising the compression, lightening the flywheels, fitting a bigger carburettor with consequent increase in port and valve size (the valves in the Venom were the same size as the MSS, but a Nimonic 80 steel exhaust valve was specified for the Venom) produced a machine with 100 mph potential. Production of a super sports roadster was an obvious move. Appropriately named the Venom, it was introduced in 1955. Alongside the Venom a 350 version, the Viper, was developed. This was in no way related to the alloy MAC but was a small-bore edition of the Venom. In fact the bottom ends of the two engines were identical and in order that a common crankcase could be used a loose collar is used to make up the difference between the size of the 350 barrel and the 500 version. The heads differ only in respect of hemisphere and port sizes. With a bore and stroke ratio of 72x86mm the Viper responded even better to the tuning technique, producing 27 bhp at 7,000 rpm, which serves to substantiate what Charles Udall once said when quizzed about the merits of long and short strokes. He could, he said, get the same characteristics from either. The Venom in road trim was good for 100 mph, the Viper for 90 mph They opened

---

First and last Thruxton

IT DOESN'T look like a Thruxton or even a Venom. More like a special some mature enthusiast has put together for serious long distance touring. The black no-nonsense finish, the touring handlebars and the army type canvas panniers do not suggest a sports model. The long-distance touring bit is actually correct for owner David Ward, an assistant curator at the Stanford Hall Museum, bought this 'one off' machine for that very purpose and has had no cause to regret his purchase back in 1971. It is special too for it was the Veloce works development machine which began life as a Venom which Sales Manager - Director Bertie Goodman used to thrash about the countryside. Probably the one referred to by his wife Maureen earlier when she complained that he took her to a club dinner dance in a sidecar and turned it over on the way home. Its original registration as 163 KOJ dates to September 1963 with engine VM 5085, but some time later this was changed to VMT 101 and became the development model.

It has the go faster Thruxton goodies of big valve head, 9-1 comp ratio,

Dave Ward's one-off Velo — 'a real piece of history'

twin leading shoe front brake, heat shield on the oil tank and a 36 mm Amal Concentric (which apparently gave the same performance as a GP with better manners and easy start-a new chapter in Velocette history. An association between Velocettes and Mitchenalls, the glass-fibre specialists followed.

Glass-fibre enclosure was developed for the lower works, which suited the contemporary fashion for clean design and enclosure and marginally improved maximum speeds. It offended the dyed-in-the-wool (or oil-impregnated) enthusiasts, but the shields were easily removed (the cost of the shieldings were offset by not having to polish the crankcase and gearbox shell, and the lack of dynamo inner and outer cover or final drive sprocket cover. One minus point was that many owners neglected to keep the gearbox oil level maintained — out of sight, out of mind). More effective was the Mitchenall Avon dolphin which, although "touring" by modern day fairing standards, added between 5 and 10 mph according to conditions. Machines thus fitted were named Venom/Viper Veelines. With slight modifications one of these fairings made a vital contribution to the successful attempt on the 24 hour world's record. The lessons gained from the record success encouraged Veloce to produce a brace of super-sporting 350/500s, with "Ace" bars, rear-set footrests and TT carburettors. Named the Viper or Venom Clubman, they added the name Veeline when fitted with the sportier Avon Veeline fairing, made to match the rear-sets and fitted with a most effective double-curvature screen.

Prestige machine of the pushrod range was the Venom Thruxton 500. It incorporated all the accumulated experience gained in the years the Venom Clubman has been raced in long-distance events, plus the development of the 12 and 24-hour record breaker. Some of the know-how that went into it dates back to the GP racing days. It is in the cylinder head that the most important modifications were made. To enable higher compression ratios than are possible with the Venom the combustion chamber was flattened and the included angle between the valves reduced. This enabled a

---

ing) but not the cafe racer cosmetics of clip-ons and humped racing seat. The reason for the sheep's clothing for this wolf is that the man who did the development mileage on the road liked a bit of comfort. That's what attracted David, who has toured widely, often with his wife on the back.

It's the special bits you see when you get close that make this unique bike so interesting. Like the 12 volt alternator where there should be a dynamo which is driven by toothed belt. In a cylindrical housing that looks a bit like a dynamo is an epicyclic reduction gear to step the drive down to an auto advance coil ignition contact breaker. Just in case this gave trouble there is another car type auto advance contact-breaker in the original magneto position. You can swap from one to the other by changing a connector.

The benefit of the 12 volt electrics is a brilliant headlamp and instant starting. I've seen it started first kick after lying unused for several months. Unseen developments which Velo men might have enjoyed if production had continued are "O" rings on leaky bits like the kick-start shaft and a crafty plastic seal between the primary chain case and the gearbox instead of a piece of cork. The rev-counter, by the way, takes its drive from the alternator shaft up front which makes a neater cable run and no oil leak.

A real piece of history and one of a batch of works machines including the Roarer, the Model O and a 250 LE which the late John Griffith, co-founder of the Stanford Museum, bought in the nick of time before the works closed.

Alternator/contact-breaker is unobtrusive

Left: The Venom and Thruxton models were used by clubmen during the 60s and 70s for racing and high-speed trials. Pete Walters negotiates the Gooseneck at Cadwell Park on his well-modified 1963 Venom during a Vincent Owners Club High Speed Trial meeting. Note the raised silencer, a Geoff Dodkin modification

2in inlet valve to be used without fear of it hitting the piston if the engine was accidentally over revved. The inlet valve is a beautiful bit of work with the shape and delicate grace of the base of a champagne glass. To aid breathing the valve seat is radiused and the inlet port has steeper downdraught and a greater swirl angle. The carburettor was a 1 3/8in GP Amal mounted on a finned spacer to provide the optimum length of inlet tract. A Venom cam was used but the followers were modified to provide an improved timing pattern. This was done by both altering the radius of the follower pad and the point from which the radius is struck. These final timing modifications were finalized by actual tests at MIRA, a variety of shapes being tried. It was found that the best timing on the bench was not always the best on the track. In other respects the Thruxton resembled the Venom Clubman with added options and all the trendy "goodies".

"You know", said Bertie Goodman, "we used to have the failing of making things technically correct though not necessarily what people wanted. At the time the Thruxton was evolved I was chiefly concerned with sales and I did things the other way round. I asked dealers what they wanted. It seemed that after buying a bike the customer would then go and buy a lot of "goodies" and alter this and that. So I thought, why don't we do it for them and build the goodies - the sensible ones - in? That way they will be cheaper. So we gave them a twin-leading-shoe brake, a carburettor spacer which looked right and was the right length, and a pukka GP carburettor. Alloy rims, a racing seat, the lot."

The Thruxton tank was modelled on the late Venom model but with a cutaway for the GP carburettor, was finished in silver, as were the mudguards and the battery box. The frame was enamelled a pleasing shade of blue. A far cry from the traditional Velocette sober black lined in gold with a minimum of chrome. It was not long before enthusiasts began to ask for the traditional finish! More Thruxtons were turned out in black than in silver. The preference for black and gold, traditional colours of *the marque* with a unique history, typifies the outlook of the Velocette enthusiast. By choosing a machine which was technically old fashioned, an oddity in a trendy space age,

Prestige machine: the Venom Thruxton 500. This one appears to be a pre-production prototype; later, the rear mudguard stay loop was never bridged by the rear number plate. Another Velocette red herring!

they thumbed their noses at progress. The lusty single Velocette may be a symbol of a bye-gone age but it was the golden age of motorcycling and there are many who wished to perpetuate it by scorning the flamboyancy, frenzied followers of fashion. I have on occasion referred to Velocettes as the only manufacturers of vintage machines. Neither the firm nor their customers took umbrage. Rather they take it as a compliment, which is how I meant it.

It takes courage to drop out from the rat-race of multis and a different colour scheme each year and chrome with everything. Being different from the rest binds Velocette owners into a tight-knit cult. The characteristic weaknesses of the single-cylinder design, the ritual starting, the low-periodicity vibration which is not felt by the rider but is death to mudguard stays and oil tank lugs, and the mysterious clutch provide an initiation to the cult which effectively sorts out the men from the boys.

The family spirt of a firm like Velocette spread much further than Hall Green. It spread to countless Velocette agents up and down the country and overseas who in time became so dedicated that they stuck to the marque through thick and thin, sometimes forswearing other makes which by reason of changing fashions, cut prices and big production might have made them more money. Inevitably when a dealer became a Velocette agent he became a Velocette enthusiast, either as a rider or as an entrant — the word sponsor is of recent origin in this sense — and inevitably on some occasion he would call at Hall Green to find not moguls but other enthusiasts, from the Goodman brothers downwards. As you listen to now-venerable dealers nostalgic stories you note that they speak with warm family affection of Mr. Percy and Mr. Eugene and Miss Ethel, and inescapably the Managing Director was "young Bertie" and the Works Manager was "young Peter". For they are talking of days an age ago although time stood still a great deal more at Hall Green than at most places I can think of.

Putting Britain back in the records book

"Fate sometimes rewards a trier"

Throughout the history of motorcycle record-breaking there have always been natural landmarks of time and distance. The most significant of these have always been associated with the magic 100 mph figure — the first machines to do 100 mph briefly in various capacity classes. Then the first machines to average 100 mph for longer periods. The classic hour, 12 hours and 24 hours. The immortal ohc Velocette had carved its niche in the list of century breakers as long ago as 1928 when Harold Willis, Veloce director, designer, tuner and rider had lifted the 350 cc hour record to 100.39 mph and the record-breaking sortie of that year had ended with the hour, 100-mile and 1,000 mile records standing to a Velocette, so that the firm could proudly advertise "The Fastest 350 in the World."

These achievements were obtained on but slightly-modified though highly tuned production machines, for the reason that in those days most manufacturers, and Veloce Ltd. in particular, raced what they sold and sold what they raced. The gap between production roadsters and grand prix racers had not opened out to a chasm which even ad. men are "pushed" to bridge, these days.

Gradually record-breaking became a specialized publicity-seeking enterprise for which only modified grand prix or "one off" machines were suitable.

Against this background, it seemed unlikely, in the 50s or 60s, that is, that any worthwhile world speed record, and certainly a record with the magic "ton" status, would ever be taken again by a modified roadster, let alone an "old-fashioned" single-cylinder. It was, therefore, almost beyond belief when a 500 cc Velocette Venom, virtually the same as any private owner could buy and so modify, proceeded to take the world 12-hour and 24-hour records for all classes at Montlhery in March, 1961. Even more remarkable, the Venom had at long last broken the century barrier for a 24-hour run, a "first" which no subsequent achievement can ever surplant in the annals of record-

The Venom record-breaker at Montlhery. Bertie Goodman takes over for an hour's stint

breaking. The fact that the Velocette still holds the 24-hour record for 500 cc machines at 100.05 mph is a continuing tribute to the magnitude of its achievement in 1961.

Yet the achievement, which put Great Britain back in the record book in a unique position, when in other classes our once proud position was being whittled away, was so much a "shoe string" enterprise, a part-private, part-works tie-up, as to be beyond belief when considered in retrospect.

Starting with the machine, it was a rather second hand development bike, prototype of the subsequent Venom Clubman Veeline. The engine had been bench-tested and track-tested at MIRA for 1,400 miles at over 100 mph the preceding August, but *had not* been dismantled afterwards. It was not, in fact, dismantled again until the record had been taken. When then dismantled for measurement, it stripped in perfect condition.

It ran on 94-octane Esso fuel for the simple reason that 100-octane fuel was not then available in France! The oil was a 20/40 multigrade mineral and was not changed during the run. The track was lit at night by nothing more sophisticated than 52 Marchal car headlamps; the hypnotic effect of these lamps nearly ruined the venture, as will be explained later. In the motley Anglo-French team of riders only two of the Frenchmen, 55-year-old George Monneret and his son, Pierre, could be said to be really experienced at banked track record-breaking. Bertie Goodman even did his hour stint at night without having practised in the dark at all. Some of the riders just proved incapable of lapping at the required speed. Pit work was carried out with quite primitive equipment, refuelling with a tin funnel and a plastic bucket! Yet the attempt succeeded beyond expectations.

It all stemmed from an approach from Georges Monneret to Velocettes. Georges was France's Grand Old Man of motorcycle racing and record-breaking, a national hero who had been followed by his equally brilliant son, Pierre. They were old hands at long-distance record-breaking, and Georges had past associations with the Velocette *marque*. In Bertie Goodman he found an enthusiastic supporter for the rather surprising suggestion that world records could be taken with a near-standard roadster single.

"In fact," said Bertie, "I didn't think the records, the 24-hour one, at any rate, could be taken with a racing machine. A racing machine is not designed or developed to run for such distances without attention." Which makes sense if you follow the dictum that the perfect racing machine should be designed to finish the race and no more. Not perhaps so developed that it would blow up as it crossed the line, but there is obviously no point in designing a grand prix machine which will run flat out for 24 hours without a spanner on it. The idea appealed to Bertie because it was in the old Velocette tradition of co-operation between private runners and the factory. It also appealed, I am sure, because it offered the possibility of some high-speed riding for him... the development testing, and perhaps a ride in the record attempt.

"Breaking the 24-hour record did not appeal to me very much... it only stood at about 85 mph then [before the Velocette attempt it had been raised to 96.42 mph by a French team on a 500 cc BMW], but I thought we might be able to top the 100 mph mark, and that would be really worthwhile," said Bertie.

"The development problem was to get the necessary life out of normally replaceable items. The primary chain was a case in point. It had to last for over 2,000 miles under flat-out conditions. Altogether I did over 2,000 miles at MIRA getting things right."

The primary chain problem was solved by reverting to

AN AVERAGE OF OVER 100 m.p.h. FOR 24 HOURS FOR THE FIRST TIME

Velocette

Riding a standard production **VENOM CLUBMAN VEELINE** a team of eight riders broke the world record for 12 and 24 hours for 500 c.c., 750 c.c. and unlimited c.c. solo motorcycles.

12 hours at an average speed of 104·66 m.p.h.
24 hours at an average speed of 100·05 m.p.h.
(Results subject to official confirmation.)

RIDERS
Bruce MAIN-SMITH   B. J. GOODMAN
Andre JACQUIER-BRET   Georges MONNERET
Alain DAGAN   Pierre MONNERET
Pierre CHERRIER   Robert LECONTE

This magnificent achievement proves Velocette performance and reliability under the toughest conditions.

FOR SUPERLATIVE PERFORMANCE YOU CAN'T DO BETTER THAN
Get a VELOCETTE

Post the coupon now for 1961 literature

To: Veloce Limited, York Road, Hall Green, Birmingham, 28.
Please send me Catalogue.
NAME......................
ADDRESS...................

A27

"We did it!" As advertised in the weekly magazines

racing practice; an exposed chain (save for a racing-type alloy guard) and a Reynolds-type twin oil feed directed on to the side plates and fed from the oil tank. The feed was taken from a point well up on the tank so that the chain oiler would not run the engine dry.

A big factor in the final success was the contribution of the Avon Veeline dolphin fairing. This was evolved by cutting and shutting a touring fairing until it not only added the best part of 10 mph to the speed but gave the riders valuable protection.

"We got the Perspex developed to the point where it kept perfectly clear in the rain... raindrops were just blown off," he said.

The machine was developed to the point when it would bomb around a track seemingly for ever at around 106 mph — bomb is the right word for the one concession made towards speed by the Montlhery regulations was freedom to use any type of exhaust system, and Velocettes used a KTT-type megaphone with a 4¼in mouth. With a 1 3/16in GP Amal carburettor on the inlet and a 34in exhaust pipe on the outlet, the valve timing employed was inlet opened 55 degrees before tdc, closed 65 degrees after bdc, exhaust opened 75 degrees before bdc, closed 45 degrees after tdc Which is exactly the same timing as the Mk VIII Velocette! Peak torque and peak power, 39.8 bhp were both between 5,800 and 5,900 rpm.

But despite all the preparation and testing it was the completely unexpected which in the event threatened to jeopardize the attempt. Not a mechanical problem but a psychological one, a failure of the human senses to adjust to the unnatural conditions of a pattern of lights in a darkened bowl. The daytime run, starting at just after 8 am, had gone off well, the Velo lapping at 110-112 mph, the average remaining at over 105 mph despite changes of rider and the rear tyre. When darkness fell Alain Dagan and Pierre Monneret kept the schedule up and the 12-hour record was taken at 104.66 mph (the old record of 102.3 mph was set up by Fergus Anderson and Bill Lomas on a 350 cc Moto Guzzi racer in 1955). Both tyres and the rear chain were changed and the foot-change mechanism, which had been strained by some heavy-footed rider, was rectified by Jack Passant in 33 minutes before Bruce Main-Smith set out. He was the first to succumb to the hypnotic conditions.

As he wrote at the time, after describing the gruelling of bumps which deflected the forks to the limit: "The next

---

Jack Passant

JACK PASSANT worked at Hall Green for 18 years and though he came to be well known as a race mechanic the "glory days", as he calls the days of Velocettes in ascendency, were pretty well over when he went there as a race mechanic. He spent most of his time testing and developing engines for road work but some work was done, after the firm stopped racing, for dealers who carried on racing, like Arthur Taylor, Reg Dearden, Geoff Dodkin and Ralph Seymour. His moment of glory, in my opinion, was his calm spanner work during the 24-hour record run which undoubtedly saved the day. Heavy footed French riders had cranked the 'bent tin' late type link between the pedal and the selector, so that the gears would not engage properly. Fortunately they had a complete gearbox end cover with the old type mechanism, and this was substituted. He agrees that they were very lucky with the 500 and equally unlucky with the Viper attempt, a year later. The same bike was used but the fairing had been stiffened at the front mounting which reduced airflow (you can identify the Viper because it wore a large Velocette logo on the side of the fairing, whereas the record breaker was plain).

Passant says the Viper was running so hot he did not think it would have lasted the distance even if they had not had to change the magneto which meant lifting the barrel and replacing a broken cylinder head stud.

"Unfortunately they were using a sand cast piston instead of a die cast one, which was a mistake, and the top came off. But it was overheating badly", he says.

Of all the products he worked on and tested he is most scathing about the Viceroy. "Fancy making a machine where to get the back wheel out you had to lay it on its side and then take the silencer off. The engine was based on a Johnson marine engine. Reed valves were ok on that because it was a fixed speed engine. A scooter is not a constant speed machine," he says.

Maybe that's why users of the Velo engine in fixed-speed devices like hovercraft all seem to have been satisfied.

problem is the sameness of the course at all points. The rhythm of one's fast, regular, monotonous progress. I realize I am getting hypnotized by the pattern of what I see. I look at the Velocette, the fairing, the red lanterns, the illuminated scoreboard and the stars, always coming back speedily to the yellow line 3ft to my left under the spinning front wheel. After 52 minutes, some 60 laps, that is, I know it would sabotage the attempts if I continued."

The next rider, a Frenchman, could just not make the schedule speed in the dark and had to be taken off; the next Frenchman lasted but 35 laps. Pierre Monneret restored the situation with a full stint but Pierre Cherrier soon came in complaining of fog on the course after dropping to a crawl. Even Georges Monneret succumbed after half an hour, complaining of blurred vision. Bruce Main-Smith managed another half-hour shift, and then with great courage Bertie Goodman set off into what was for him the unknown. He stuck it out for a full hour, only 2 seconds a lap slower than the average daylight time.

This is how he recalled his experience in that longest night:

"At night we had trouble with successive riders starting off OK and then slowing and coming in complaining of fog on the track. There was no fog, of course. Yet Pierre Monneret had no complaints. I set off like mad at first. There was no particular problem at first but after 20 minutes I found I did not know where I was. I was completely disorientated. It was so unreal. Down the straight it was like a black tunnel. Then at each end there were 21 lights... plonk, plonk, plonk they went. We were becoming hypnotized by the pulsations of the light — the same effect as a psychedelic flashing light. My flight training helped me to get my bearings again. I sat up so that I could see the skyline above the banking... lying down to it was like riding in a cup unable to see the rim. I did this on each lap to get my bearings and take my eyes away from those pulsating lights. It cost 0.1 or 0.6 second per lap but I was able to keep going."

The rest of the record run was eventful. Dagan had no trouble with the darkness and at daylight, Main-Smith and Bertie Goodman had another go, and to Georges Monneret was accorded the honour of doing the final shift. Alas, he had to give up with frozen hands and Dagan, the indomitable, finished the job.

One day, no doubt, another team with another machine, this time perhaps a "modern" multi with power to spare (the Velocette was flat out all the time) and perhaps backed by a team of star riders and attendants, may wrest the 24-hour title from Velocettes, but no future achievement can ever dim the glory of being the first to reach this historic landmark. Nor is it likely that there will be many bids for the record for there can be so many slips between cup and lip in a full 24 hours, and to a prestige-preoccupied manufacturer the price of failure weighs heavily against the possible reward of success. For Velocettes, and Bertie Goodman in particular, the record attempt was a case of "having a go" in the old sporting tradition of the firm. Luck, which so often decides the outcome of such ventures despite the most meticulous preparation, was with them on this occasion.

The records (for 500, 750 and 1,000 classes) taken at Montlhery on March 18-19, 1961 were:

12 hour: 2,021,181 km., 168.431 kph, 104.66 mph
24 hour: 3,864,223 km., 161.009 kph, 100.05 mph.

The riders were: Pierre Cherrier, Alain Dagan, Bertie Goodman, André Jacquier-Bret, Robert Leconte, Bruce Main-Smith, Georges Monneret and Pierre Monneret.

A few years later they "had a go" at the same records with a Viper. It was fast enough, being able to lap at 108 mph. But the first attempt was ruined through the silliest fault. The plug lead which should have been soldered to the little copper washer in the pick-up wasn't. (The magneto was one part Velocettes had not prepared.) The plug lead worked out of the pick-up and ruined the armature. They changed the magneto but in lifting the barrel to get at the fixing nuts a holding down stud sheared. It took a long time to extract and replace. So they started the attempt all over again. This time the piston (not of Velocette manufacture, Bertie emphasized) failed, after 1½ hours. That was that.

Like the Roarer and the Model O, the 24-hour record-breaker was bought by John Griffith. Subsequently it was acquired by the National Motor Cycle Museum near Birmingham where it is displayed.

Neil Kelly: last TT winner for Velocettes

WHEN the ACU announced that a Production Race was to be included in the Diamond Jubilee 1967 TT, it was greeted with enthusiasm by all the home manufacturers. The Japanese had still to make an impact in the larger classes, so it was expected that Triumph, BSA, Velocette et al would be fielding competitive machinery (*writes Bill Snelling*).

The Velocette entries were fielded by Reg Orpin of L. Stevens of Shepherds Bush, and Geoff Dodkin of East Sheen. Veloce had been working on a 'super-Thruxton' engine with a squish-band head, and one of these engines were supplied to each of the teams. Geoff Dodkin's rider was Keith Heckles, a great favourite in the Manx GP, but Reg Orpin was in trouble. His first choice rider was Dennis Craine, who had won the 1965 Lightweight MGP for Reg on an Orpin Greeves. But two weeks before the TT he had crashed in a Manx scramble and broken an arm. Coincidentally, the rider who was involved in Craine's tangle was 30-year old Neil Kelly, who was then asked by Reg if he would ride his Velocette. (Neil assures me he had no prior knowledge of the ride when he rode over Dennis's recumbent body!). Standard Thruxtons were ridden by Arthur Lavington and Bob Biscardine.

The race bike was late arriving on the Island, so Neil was forced to practise on an MSS, and a dog rough one at that. It was plagued with all sorts of mishaps from the first practice session, and come Thursday afternoon, the last practice session for the roadsters, and Neil had not completed a single lap; he had to get the bike round *somehow*. But it was not to be. The clutch started slipping just after Ballaugh and he finally pulled up at Quarry Bends. A spectator there must have been a Velo rider. He promptly pulled a nail out of a fence and adjusted the clutch for Neil! Just as they were about to set off, the roads-open car came past. Neil did not have a single lap to his name....

Thinking that his chances had gone, he went to work the following Saturday, but his local friends had been badgering Mary Driver, the then Secretary of the meeting, to let Neil ride: he had six years' Manx GP experience under his belt. There was a frantic phone call to his place of work that Saturday lunch time: if Neil could get to the start he would have his ride! He recalls getting his boss to run him to the Grandstand. Reg had already weighed the bike in. His leathers were collected and all he had to do was turn up and ride. All, that is, except start the damn thing! Neil had spent some time practising the mysterious art of getting a Velo fired up, but when the flag dropped and the 500 class riders scuttled across the Glencrutchery Road, he was nearly last away. But Neil was not alone

Neil Kelly takes the chequered flag to win the inaugural Production TT 500 class

sweating, swearing and kicking, Keith Heckles on the Dodkin Thruxton was also suffering, and these quick Velos were the last to leave the grid.

Once under way, Neil settled in for a quick three laps, but from such a tardy start it was difficult for him to gauge his progress, though for sure he was passing lots of riders, including slow 750s as well as any number of 500s. Percy Tait, a fancied winner on a Triumph Daytona, crashed at Appledene on the first lap and Neil pulled away not knowing where he was. The squish-head Thruxton was certainly going far better than his practice MSS.

Starting the last lap, Neil caught sight of Reg holding a board out with '1' on it. He thought it meant one lap to go; the penny hadn't dropped! It was only when a local friend nearly hit him with a board at the Hawthorn pub which read 1st + something or other that he finally twigged that he was ahead of all the 500s. His second lap was completed in 24m 52.6s, 91.01 mph, the fastest of the race.

The rest of the last lap was a worrying time: would the motor last, would it run out of fuel, all the things that flash through a rider's mind until the chequered flag falls. Keith Heckles pulled through to second place, with Dave Nixon third on a Triumph Daytona.

The post-race scrutineering was a slightly nerve-wracking affair. Reg Orpin had added some gussetting around the swinging arm, and the silencer most definitely wasn't! Silencing, that is. If you got the light right you could clearly see the megaphone that had been inserted therein. Would anyone protest? No, the result stood, and Neil went on to collect the magnificent sum of £50 prize money. Well, not strictly correct, because Reg insisted on relieving Neil of 50% of his winnings, as sponsor. Not content with that, he also took half the £50 bonus that Neil had won from the plug company!

Veloce paid Neil the compliment of a like amount as a race winning bonus, but this time Reg didn't get a share.

Neil Kelly and Stanley Woods, pictured at a TT week Velo Club meeting

Next year Neil was again mounted on a similar machine, again with a squish-head motor, but this time a dodgy magneto kept maximum revs down to 5,600 and he struggled to finish fourth in a race won by Ray Knight (Triumph) from John Blanchard on the Dodkin bike.

That Clutch...

Scourge of a new owner, very often a pain even to long-time Velo fellows

The most controversial feature about a single-cylinder Velocette is the clutch. Always has been and still is. Because it is unconventional in operation and appearance, it has always been the victim of inexpert mechanics and the butt of uninformed criticism. To this day it is a mechanism which sorts out the Velo men from the boys. Yet it is a very good clutch, conferring by reason of its slim design great benefits to the function of engine and gearbox in the reduction of overhang loads. It can in many ways be likened to a beautiful woman. It is slim but surprisingly strong. Feather-light to the touch when handled correctly but a perfect devil when mauled by uncouth hands. Capricious one moment, captivating the next. But different, so different that many people have owned Velocettes without ever really understanding the subtleties of the clutch and many have given up Velocettes because of it. (A big motorcycle dealer once told me that he never took the Velocette agency because he was scared of the after-sales service trouble he would have with customers who did not understand the clutch.)

Much of the trouble that has occurred with Velo clutches in the past is due directly to the motorcyclist's compulsion to fiddle with things. If the maker provides a knurled adjuster on the clutch cable it is regarded as an invitation to fiddle with it. With a conventional clutch one of two things happens. The clutch either slips or does not free but the result of the adjustment is fairly obvious and can be remedied. Fiddle with an adjuster on a Velocette clutch cable and the results are not immediately obvious or logical. On a Velocette the cable adjuster is only a means of initially setting up the cable, a secondary matter, and not for the main adjustment of the clutch. In fact, the beginner is best advised to forget about it and that is why Velocettes usually hide it under the tank.

I had better explain the general layout of the clutch. The clutch cable which disappears into the gearbox is connected to a bell crank in the gearbox which presses on a short pushrod which emerges from the gearbox shell behind the clutch and presses on a hinged plate. The hinged plate presses on a thrust race. This in turn, presses on three thrust pins, which run through the clutch back plate and lift the outer pressure plate against the pressure of the clutch springs.

The cause of most of the confusion is that the bell crank is inside the gearbox and can be "bottomed" at the end of its travel while the clutch cable is still showing free play. You then have the condition where the clutch slips but there is free play on the cable. Or perhaps it does not slip but the thrust race is under constant load and wears out, yet there is still slack at the cable. In the old days of motorcycles with exposed clutch arms on gearboxes the Velocette was perhaps the only one out of step with a hidden operating arm (the bell crank) and the only one, consequently, to regularly suffer from lack of intimate knowledge. As other makers gradually enclosed their operating arms they ran into the same problems, but with a conventional clutch the push-rod takes the brunt and is easily replaced.

I am not going to attempt to explain the sequence of adjustment. The maker's instruction book devotes pages to the subject and although it may sound a lot of mumbo jumbo I can assure you that if followed blindly, without even trying to understand the whys and wherefores, it will work. The trouble is that if you have had, like most people, an apprenticeship on conventional clutches you will be tempted to skip the marker's roundabout instructions and try a short cut based on your previous practices. Don't do it.

But human nature being what it is people still try to be clever and of course you may find yourself without an instruction book. For such cases I have, after thought and personal experience, evolved a simple man's (a man used to normal clutches) conversion guide to the mysteries.

After considered thought, I came to the startling conclusion that there is nothing very unconventional about the Velocette clutch. It's just that the parts look different and tend to be in different places. For instance. The conventional clutch has a long pushrod which goes right through the main shaft. The Velocette has one in the gearbox shell which transmits the thrust to the thrust race (conventional clutches don't have the sophistication of a thrust race — they just wear out their push rods) and three little pushrods which run through the clutch. None of these pushrods rotate and they do not wear out.

The average pushrod presses on a hardened button or ball in the centre of the outer clutch plate and if it is a sensible design there will be an adjustment for the button or

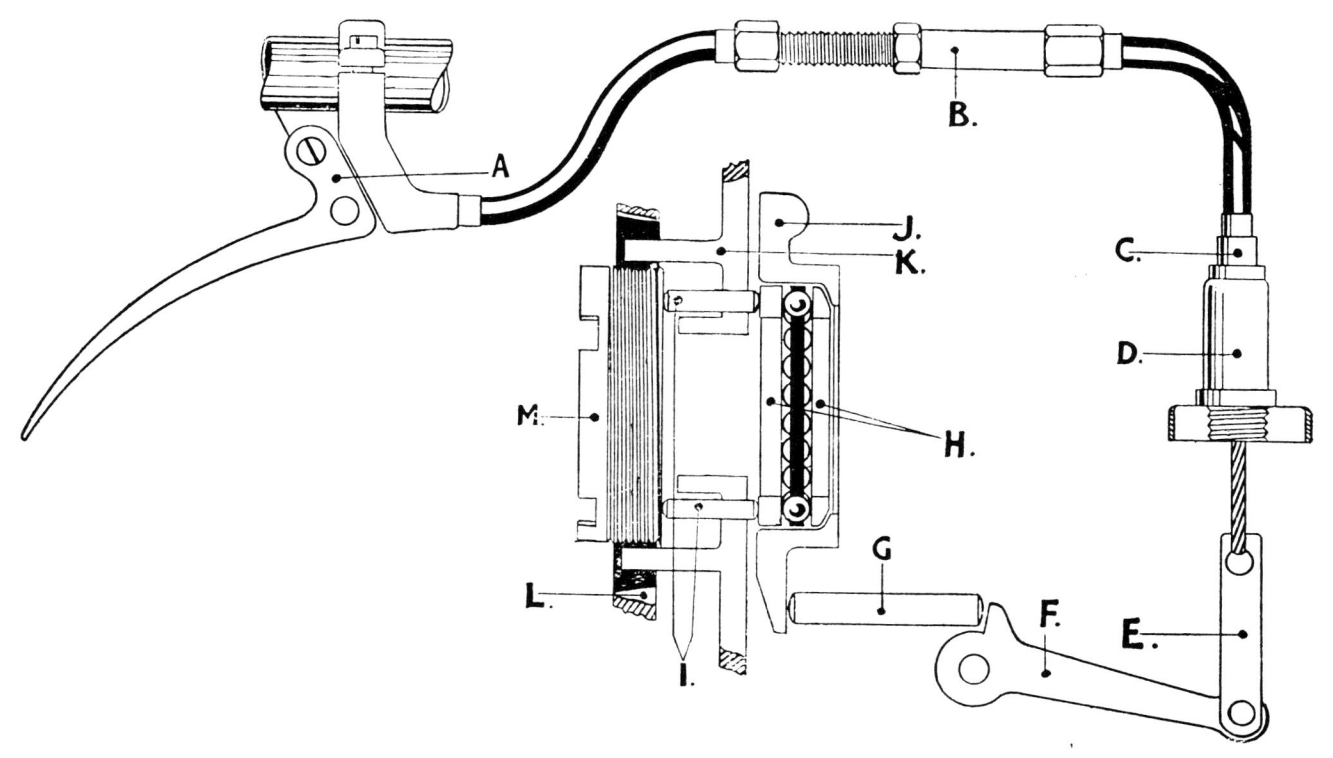

Clutch-operating mechanism. A: handlebar lever, B: cable adjuster, C: cable stop, D: cable-stop holder (on gearbox), E: cable connecting piece (in gearbox), F: operating lever (in gearbox), G: large thrust pin, H: thrust race (three parts), I: thrust pins (in back plate), J: thrust cup, K: back plate of clutch, L: front plate of clutch, M: spring holder.

ball which you should use when you have run out of adjustment at the cable-operating arm. Instead of a button or ball the Velocette has a ring screwed into the outer pressure plate on which all three push-rods seat and this ring — it's the one with castellated slots, provides the main adjustment for the clutch operation. But in the manner of its working, it is exactly the same as that adjustable button in the outer plate of the normal clutch. Screw it in for less play, out for more. There is no adjustment for spring tension and none is necessary for with the sixteen or more small springs, instead of perhaps six big ones, individual differences between springs are unimportant. The ring which retains them and has holes in it for a peg spanner (please, not a nail) also acts as the nut which holds the clutch on, and wants to be tight.

The only real problem once you have related the Velocette clutch to the conventional clutch in this way is that you cannot see, or feel, with chain guards in place, the vital free movement necessary at the thrust race, and the cable is no criterion at all. That is why it is necessary for the instruction book to take you through the mysterious route of first slacking the cable right off and adjusting the clutch thrust sleeve and then adjusting the clutch thrust sleeve (the equivalent of the thrust button, though the makers called it the clutch spring carrier — which it is, incidentally, until the clutch slips slightly on the kick start.) This to ensure that the bell crank in the box is at the end of its travel. The cable is then adjusted until there is no free play and then left severely alone. The clutch thrust sleeve — I think my term is less confusing — is then adjusted until the cable develops 3/16in free movement, and all should be well.

Provided that the thrust race and plates are in good condition. If the thrust race parts or clutch plates are worn there is no real choice between a slipping clutch and a dragging clutch, for if the Velocette design has a fault it is there is not the wide range of adjustment provided by other designs to enable you to go on using a worn-out unit.

You may wonder how it is that a clutch so slim and puny, with a handful of springs like those out of a petrol lighter, can transmit so much horse-power with such a light action on the clutch lever. Very much, in fact, like the latest diaphragm clutches. Well, the explanation as I see it is that the Velocette clutch after 1926 has actually a servo operation. The action of the hinged lever at the back of the clutch is to lift one side of the thrust race only. This lifts *half* the clutch only, the plates tipping sideways like a conventional clutch which has incorrectly adjusted springs. Rotation of the clutch by a half revolution lifts the other side; the spherical seating behind the clutch thrust race allows the race to line itself up and the clutch frees (or should do). You can watch all this happening if you watch the clutch while you lift the clutch lever, and although with the lever fully lifted the clutch may be in slipping condition, because some of the pressure is taken off the side which has not lifted (the amount of tilt of the plates depends on the fit of the sprocket on the sleeve gear extension), it will not lift fully until it has been rotated.

This is where the servo bit comes in. As you already have exerted your operating force in lifting half the clutch, the remaining force required to lift the remainder must come from the machine itself in the rotation of the clutch - movement of the machine, if the engine is not running. Hence the light and disproportionate feel of the clutch lever.

Quite frankly, I stumbled on this servo theory when studying the action of the clutch for the purpose of this chapter, and when I propounded it to various Velo experts and even to head men at the works they were at first politely unbelieving. Certainly it has never been put forward as one of the advantages of the unusual form of operation which went into production in around 1925 and never altered (the number of plain and inserted plates and springs has increased through necessity of transmitting more horse power). It should be noted that the first clutch used on two-strokes immediately after the first world war did not have this servo action, for the pressure plate was lifted by a quick-thread mechanism pressing axially on the pressure plate *via* the traditional three thrust pins, and this design is to be found on every three-speed Scott from 1927, which is about the period when Velocettes abandoned it.

I wonder if Percy Goodman adopted the hinged thrust mechanism for the ohc models to obtain this measure of servo operation when spring pressure had to be increased, or whether it was an accidental bonus. If he did it deliberately, and he was too good an engineer to overlook a principle like this, he kept very quiet about it.

If you are still unbelieving you can quite easily test the theory for yourself. With the primary chain removed and preferably some of the clutch springs, so that the operation become easy, lift the clutch and then turn it through 180 degrees until it frees. You will distinctly feel the resistance to rotation as the clutch lines itself up. This extra force must come from the machine, not your hand on the clutch.

The last all-new design from Hall Green
Viceroy scooter was a final - fatal - throw at the 'everyman' market

Although you may frown at mention of a scooter, this Saga cannot be completed without analysis of the 250 cc Viceroy scooter introduced in 1960 and manufactured until 1963. Although around 1,000 were sold, this was such a drop in the ocean in the scooter world that they never became a familiar sight on the road and soon after became virtually extinct. Yet unlike the many other British-made scooters which blossomed briefly and then vanished without trace in the same period, the engine unit of the Velocette Viceroy lived on, having found favour in a completely different medium where its unique qualities found more appreciative users.

The real importance of the Viceroy in the story of the Velocette *marque* is that it was the last completely new design to come from Hall Green and the drawing board of Charles Udall. Technically it was perhaps the most correct concept for a luxury scooter to be made in this country or anywhere else; price-wise it was, considering its specification, a "best buy" yet prospective scooter buyers fought shy of it. The failure of the Viceroy to sell in appreciable numbers is an object lesson to anyone concerned with the design and marketing of consumer durables to a sheep-like mass market.

One may wonder why Velocettes, having only just got over the disappointment of the LE, should have risked another throw at the fickle, near frenetic, "man in the street" market.

"Why?" I asked Bertie Goodman. "Was it pressure from dealers?" I remembered, you see, that several Velocette dealers, seeing the march of the scooter invasion and the rout of the LE from the commuter market, had begged for a scooter version of the LE.

"Well," he said, "we had been asked if we could make the LE more like a scooter but it was not really possible to redesign it with the flat floor the scooter customers wanted. The gearbox would have made too big a hump. And it would have been too expensive. But we felt that the British industry had not taken enough notice of the growing imports of foreign scooters and we thought we ought to try and make a good British scooter."

What they did make was what one might really have expected from a firm of motorcycle enthusiasts. It was not so much a scooter in the feminine scooter idiom as an open-framed motorcycle enveloped in pressed-steel bodywork. A scooter for motorcyclists by motorcyclists, but by and large motorcyclists looked on scooters with derision, so even dyed-in-the-wool Velocette enthusiasts looked at it with an "O Gor' blimey-what-next?" outlook. In defence of the Velocette decision to make a scooter, it should be realized that at the time of its conception motorcycles were becoming more and more smoothed out and enclosed. Rear end "bath tub" enclosure was all the rage, following the Triumph lead, and the front fairing was completing the enclosure theme. Germany had produced some elaborate heavyweight scooters of teutonic masculinity to combat the Italian lightweights and thoughtful observers, myself included, forecast that the scooter would grow more like the motorcycle and the motorcycle more like the scooter till the twain would meet in the final development of an enclosed two-wheeler.

No one foresaw that in a few years the pendulum of fashion would swing to the extreme of the out-and-out

Velocette Viceroy in 1963. In their press handout pictures the factory inclined to sylvan settings. Girls, when they occasionally appeared, could be taken for motorcyclists...

sportster often naked and unashamed.

What Charles Udall produced to answer the specification laid down by the Board was, apart from tele forks derived from the LE, a brilliant example of new thought and the use of design instead of cheese-paring, to keep cost down while retaining a lavish specification. The engine was the biggest breakaway. Velocette tradition demanded that the engine be vibrationless, which to their way of thinking meant a flat-twin. Insistence on handling to motorcycle standards determined that the engine weight be kept well forward. To provide a reasonably flat floor the gearbox and transmission had to be rearward. The basic layout was therefore determined by these considerations. To provide a 250 cc flat-twin within the budget, Udall turned to a two-stroke, an opposed "boxer" engine with both cylinders firing together and towards each other. Surprisingly, this form of two-stroke had been ignored by motorcycle designers since around 1919 when a tiny flat-twin two-stroke machine called the Economic was made briefly at Eynsford in Kent. Made rather than designed, for the engines were ex-US Army surplus "chore horses", (an example can be seen in the Stanford Hall Museum). It was the obvious way to make a flat-twin two-stroke for one compact crankcase serves both cylinders and there are no internal sealing problems. The price of the simplicity is a twin which sounds like a single, has in fact two small explosions instead of one big one and, of course, a single's torque. It met the age-old Velocette requirement of a narrow, compact crank assembly and into it Udall put his well-tried, though still uncopied, taper-roller races.

Early days. Colin Clarke adjusts the ignition on the Viceroy engine which provided thrust for the 'Clarkcushion', a pioneer hovercraft

Instead of ports for induction he took a leaf out of the American outboard motor manufacturers' book and used reed valves.

The advantages of the valves, which open under negative pressure — when the time is right, induction wise, and not a given piston position — are easy starting, good slow running and economy in fuel, despite in this case running on a Viper-size 1 1/16in Monobloc carburettor. A very sophisticated power unit, light in weight and low in cost. At the front of the crankcase was a 12v alternator and contact breaker. At the rear, a large flywheel with shrunk-on teeth to mesh with a car-type starter motor. To the boss of the flywheel was bolted the forward end of the car-type propeller shaft, rubber bushes being interposed. The propeller shaft was quite a giggle in its simplicity. The hollow tube was towards each end and splayed out like a banana to produce a two-legged spider. This delightfully simple idea worked perfectly well — only one breakage can be recalled. The rearward end of the prop shaft drove a transverse reduction, a duplex chain, to the clutch sprocket. This reduction chain drive, completely enclosed with the clutch in a separate casing, carried the drive line over to the offside of the machine to clear the rear wheel and at the same time gave the advantage of a reduced-speed gearbox, a typically Velocette four-speed positive-stop mechanism. The gearbox case was extended rearward again to house a spiral-bevel final drive, the rear wheel being mounted on an extension of the crown-wheel shaft. The complete transmission housing, therefore, acted as a single-sided pivoted rear fork swinging from lugs on its forward end and reacting against a spring shock absorber unit angled up to the rear of the frame backbone.

The frame consisted mainly of a 2½in tube bent in the form of a wide vee, the steering head welded to the front tip and the rear spring unit attached to the rear. The engine unit was hung from a point half-way along the forward slope and the transmission unit pivoted from the base of the vee. In the interests of weight distribution, the fuel was carried up forward — 2¼ gallons of pre-mixed petroil. The forks, as mentioned previously, were of cut-down LE type; Velocettes had experimented with a single-sided fork, as fitted to several other scooters, but found it wanting by their standard of steering. To achieve their standard they fitted really big wheels mounting 4in x 12in tyres. Everything was on a big scale compared with ordinary scooters and by the time they had enveloped the chassis in curvaceous panels to give weather protection at high speeds (this time

Prototype twin Viceroy- engined Hoverbat

they made full provision for an optional windscreen) the complete scooter was massive. The usual bikini-clad starlets posed with Italian scooters would have been completely dwarfed by the Viceroy. Mae West would have been a trifle overshadowed. There, I think, in a nutshell, was the prime reason for the failure of the Viceroy to capture more than a crumb of the market. It was so big it frightened timid scooterists out of their tiny minds. I have not ridden a Viceroy and the contemporary road tests of scooters are too eulogistic to be taken seriously. But I have consulted a friend who has owned two at various times and am content to be guided by what he has to say.. It was this:

"They certainly were big. It was a job to get one down a house entry. Bearing in mind the supposedly vibrationless flat-twin engine I was very disappointed with the vibration and roughness when it started up or was ticking over. It was beautifully smooth when it got going and had a lot of punch. It was best cruising at 40 to 50 mph and would steam up long hills in a very encouraging manner. The steering and handling was as good as a good motorcycle. It was more noisy than I expected, there was a rattle from the engine which I put down to the reed valves, and the chain reduction gear could be heard all the time. And the starter motor sounded like a Ford. Accessibility was good in parts and bad in others. It was easy to get at the engine, but there were too many little nuts and bolts holding the rear panel on. And a silly thing was that you had to remove the body of the silencer to get the back wheel out. It was usually stuck fast, too."

I think that is a fair assessment of the Viceroy. It was very good value for money, though. When announced it was £198, which was not out of the way for a de-luxe scooter, and by 1963 the price had dropped to £173. At that time the 250 cc twin Triumph Tigress was £198, a 150 cc GS Vespa £196, a Diana £193 and an NSU Prima £229. A 175 cc Lambretta was £189. But price in those days was not so important as prejudice. The scooter public was by this time so sold on Italian scooters that they were prejudiced against British ones and apart from the size, the odd noises, the lack of plated gew-gaws, the name, Viceroy, was an unhappy choice. It didn't go with coffee bars and the symbolism of blondes and bikinis.

The happy sequel to this story is that the Viceroy engine took to flying — well, sort of. It so happened that its virtues of low initial cost, light weight and good performance allied with reliability were just what lightweight hovercraft constructors were looking for. At first they cannibalized Viceroy scooters — that is one reason why they became extinct — and then persuaded Veloce to make engines for them. Charles Udall was recalled from retirement to do a bit more development with this outlet in mind and the future looked rosy for this ingenious little unit.

Yes, the Viceroy did fly though not very high and not very far. For a time it became the favourite power plant for the hovercraft amphibian vehicles which provided a new outlet for DIY enthusiasts in the late fifties and early sixties. The hobby turned into a sport when constructors turned to racing as an outlet for their enthusiasm and in this form it was still going strong in the early nineties. But not with Velocette engines.

One of the pioneers was Colin Clarke, a college technician whose early machine using an Ariel Arrow engine for lift and a Viceroy engine for propulsion is preserved by the Hovercraft Society as a historic specimen.

"Engines were a problem at first", he told me, "we used Villiers 197 cc engines with Dyna start electrics and

Viceroy engine in another element. Prop. powered speed boat makes waves

Clinton engines out of karts but they were not powerful enough and vibrated. The Viceroy engine seemed ideal for the job. It still was not powerful enough but it was a lot better.

It was light and because it did not vibrate, it only required light mountings. Best of all, it looked like a little aircraft engine. You could dispense with the flywheel by fitting a propeller in its place although some people kept the flywheel and self starter."

A firm at Crowland in Lincolnshire, Hover-Air Ltd., began to manufacture quite sophisticated hovercraft, offering them complete or in kit form. They offered the choice of Villiers 250 cc engines with Siba self starters or Velocette Viceroy engines complete with starter motors. Their top of the range model, the Hoverbat, described as an Air Cushion Vehicle, used three Viceroy engines, one for lift and two for propulsion. Said to be the first ACV complete construction kit in the world and easily built without special tools in one's garage, it was able to fly/hover/float over farmland, beaches, golf courses, lakes, etc. The future looked bright for the firm and for Velocettes but alas despite lots of media publicity (with starlets) the idea fizzled out and remained an enthusiasts DIY outlet with competition generating mechanical ingenuity.

Eventually Hover-Air Ltd. went into liquidation. An engineer who worked for them thought they had about 100 Viceroy engines.

"Considering it was a rather hostile environment, dusty or often salt-laden spray, the engines stood up very well. In fact we had more trouble with the Lucas starter motors than anything. But they did not have enough power, really. I remember Velocettes once provided three engines they had tuned up a bit, claiming they would give 14 bhp, and wanted us to test them out but they did not seem much different to the standard ones."

Colin Clarke said he tried to tune his Viceroy engine, which he got from a scrapped scooter, by strengthening the reed valves but it did not make much difference... merely made it harder to start.

He moved on like most of the enthusiasts, to more potent motorcycle engines, usually the alloy Triumph twin to which he added a flywheel and self starter from a Triumph 1300 car. "You needed an electric start because you could not very well step out and restart an engine with a cord when you were over water". Colin recalled that he went into the hovercraft game because he really wanted to fly but could not afford the cost. He gave up when hovercraft became too expensive with race competitors paying anything up to £1,000 for foreign twin two-strokes producing 40 to 50 bhp.

It was an outlet for the Viceroy engine which may have raised Veloce hopes, only to dash them again, like so many ventures.

The Viceroy engine got a little nearer flying when Bob Brearley and David Smith (remembered in the motorcycle world for their glass fibre tanks and fairings) moved from work on hovercraft to an air propeller propelled boat or "swamp buggy". With a wooden propeller bolted on the flywheel of an engine robbed from a derelict scooter, it performed well and gave no trouble. It seemed just the job to fit into the nose of a glider to help out between thermals but a visit to Hall Green decided them against it.

"They were friendly enough and showed us engines they were developing for milk floats or something (an ice cream venture) and two Viceroy engines arranged in cruciform for hovercraft work, but the factory was very run down. Men working machines had their own coke stoves made out of oil drums. They couldn't promise anything and we gave up the idea," David Smith recalls.

Matt Holder and the Velocette Motor Cycle Company

When Matt Holder bought the Velocette name and manufacturing rights at the liquidation sale in 1971, he was so overcome with delight that he just had to tell someone of his plans. It just so happened that Ivan Rhodes and I were the first two people he bumped into outside the factory and he promptly took us out to lunch! He was really going to make Velocettes, he had bought tooling, parts and drawings, and he couldn't talk of anything else. Many years before he had bought the Scott motorcycle in similar circumstances, made a few of the originals from parts, before bringing out a completely new up to date model. Well, not completely new because the engine and gearbox were the old type, and not many because the market for an idiosyncratic motorcycle like the Scott was always limited. But he kept on manufacturing Scott spares and then Vincent spares when he bought the Vincent manufacturing rights and Royal Enfield parts as well which had come as part of the Velocette deal.

Matt did build some Velocettes, getting on for a hundred, his son David recalls, but then the market for new classic type machines seems to have waned in favour of foreign multis. Interest in British classics did come back a few years later and looks set to continue indefinitely.

"I believe that if Velocettes could have hung on for another five years they would have been able to keep going" says David Holder who, following his father's death, manufactures and distributes spares world wide for British machines from The Velocette Motor Cycle Company, which is now housed in a building which was formerly part of the Triumph motorcycle factory at Meriden; they supply Triumph spares too.

Will the new owner of the Velocette trademark make Venoms and Thruxtons again? He could produce the major parts, no problem, but the little bought-out bits like switches and electrics would be difficult as they are no longer available. And how many could he sell?

"A lot of people tell you they would like to buy a new Velocette but it is a fantasy. The fact is that if they were offered one only a few would buy" he says. The fact is also that classic enthusiasts like to buy a bike cheap and restore it. It may not work out cheap in the end but the initial outlay is low and the expense of restoration is spread over a long period, years sometimes. One might think that in time the demand for spares for restoration would decline for the number of machines is limited and they are not usually used enough to wear them out. David Holder thinks not. Only a small proportion are, he says, restored properly and the others soon need more work while even the well-restored machines are often stored in poor conditions and eventually need to be rebuilt again. Either way, the Velocette Motor Cycle Co. wins and as he sits at his desk the phone never stops ringing from trade customers all over the world... and he knows all the Velocette part numbers off by heart even if the customer does not!

The Velocette Motor Cycle Company, Meriden Works, Birmingham Road, Allesley, Coventry, CV5 9AZ.

Velocette days

MY involvement *(writes Bill Snelling)* with the Veloce started with Stan Lewis, a family friend who campaigned a Mk VIII KTT in the early fifties. I recall going to see Stan in action at Brands Hatch; he rode at most short-circuits, but never on the Island. It was on my first visit to the TT in 1960 that I was introduced to Arthur Lavington. I worked for Arthur for a few years at his Tooting works.

I had joined the Dorking centre of the Velo Club at this time, and recall that at club nights at the Hand in Hand on Box Hill there used to be upwards of 40 Velos parked outside. I passed my test on Arthur's LE, and my first Velo was a Viceroy scooter, an incredibly complicated machine for a scooter, but one on which I rode countless thousands of miles. One run which sticks in the mind was at Easter when I was down country to watch the Land's End trial. Heavy mist, going over Porlock to Lynton — but the 12 volt lights of the Viceroy were more than a match for the murky going. Halfway across I was aware that I was playing the role of 'Pied Piper' to a whole batch of competitors who had tagged on as I passed them! This was good fun, and more so when I mistook a car park for the road; four 90-degree right-handers later and we were back where we should have been anyway!

The Viceroy's engine configuration lends itself to a two-stroke 'party-piece', the ability to run in reverse. If the engine was switched off, and then, JUST before it ceased revolving you turned it on again, more often than not it would fire up backwards. Very useful if you were ineptly partial to a bit of trail-riding and used to end up in a ditch. This happened one night when we had left the Saltbox Café at Biggin Hill and were making our way round the Surrey lanes at about 1 am. No sooner had I nosed into a bush than with a quick flick of the ignition switch and I was on my way out again. But not quick enough to notice that two colleagues had come to my assistance. The first I knew of this were the muffled curses noticing arms and legs sticking out from under the footboards!

My introduction to the MCC long-distance trials was an abortive ride in the 1964 Exeter. I had bought a Mk. II KSS-engined MAC and, following Arthur's example, geared it a touch lower and away I went down to the West Country. On Fingle Bridge, after punishing the clutch, the bush pushed out of first gear and I was left without the only gear that would pull me up Fingle's many hairpin bends.

The next MCC mount was to give me some very rewarding rides. I had tracked down a Scrambler frame, the one with the rear sub-frame based on the Mk. VIII KTT, and fitted it up with an MSS engine unit, together with a pair of the longer scrambler fork springs. A Valiant-based scrambler

● *Bill Snelling, motorcycle author, and publisher under the Amulree Publications banner of this book, has impeccable credentials as a 'clubman enthusiast'. The son of Ted Snelling, former Membership Secretary of the Velocette Owners Club, Bill went on to work with some of the best known figures in the extended Velocette family of dealer-agents, as recounted here, including of course Arthur Lavington, profiled on earlier pages.*

Currently, Velo-less, he retains memories of the many thousands of miles he rode, raced and trialed his own versions of Veloce machinery.

Stan Lewis on his Mk.1 cammy in a Vintage Rally

tank and 4.00 x 19 rear and 3.00 x 21 front tyres completed the 'Flintstone', which took me to many awards in the Land's End, Exeter and Edinburgh trials, including winning the Motor Cycle Championship outright two years on the trot, 1968 and 1969, both times with two first and a second class award. Next year I achieved a much sought-after prize, a Triple award for non-stop performance in all three trials during a year. The 'Flintstone' was in a constant state of 'development' during my trials period. In its later form it was fitted with a Criterion alternator conversion, a most worthwhile addition to any Velo used after dark. But nothing lasts for ever, and the Flintstone was sold to Adrian Pirson, former Chairman of the Velo Club, and my next MCC mount was a MDD model, with nicely situated rear-set footrests and low first gear, ideal for climbing such hills as Beggar's Roost, Simms, Bluehills Mines, Bamford Clough, etc.

Amongst a succession of Velo's of all types, one which I remember with fond memory was a green springer MAC which rejoiced in the name of 'Chuckles'; at least, this was the name stencilled on the front mudguard. Whether this was its 'grinability' factor or the chuckling sound its very worn rocker block made is open to question. The fact that a worn carburettor made nearly as much noise as the rockers, housed in an alloy sounding box, meant that 'Chuckles' was not the quietest of Velos.

After a boring year working for a car garage, I joined Geoff Dodkin's staff at East Sheen, London SW 14. This was at then new premises on the South Circular Road, I was always amused to visit his old shop in Queen's Road, when Geoff's head would appear from the cellar workshop. If a crankcase was 'on the gas'

The first hairpin of Fingle Bridge during the MCC's Exeter Trial, 1964. "I am already giving the clutch a lot of stick and am just about 400 yards and six hairpins from retiring with no first gear on the KSS-engined MAC"

to remove the outer main bearings, fumes would rise through the floorboards! I went initially as a mechanic, but later moved to the spares counter, where I spent many a happy year parcelling bits for despatch all over the world, especially Australia. The local Velo Club acted as agent for most of their members, the orders used to come to well over £200; all of which had to be parcelled up and dispatched through the local post office.

The French, oh yes, the French. They would arrive with a 'rat' Venom, which, after a week of Geoff's attention, looked like a new Thruxton. They would pay the bill, then spend the next hour rubbing road dirt all over the machine, the muckier the better. They came across the Channel on a Velo, and were returning on same, who needs bills for customs if it looked the same 'rat' bike when it went back! My French is non-existent, mostly their English was likewise, so it was a good thing that Veloce always produced well illustrated spares books!

It was during my time with Geoff that he brought the remaining stock of Veloce-engined Indian Velos. These machines were built in Italy for Floyd Clymer's Indian company. When Clymer died the remaining stock of these machines were left in limbo. Their styling offended many traditional Veloists, but they were superb to ride, even if the saddle height was a bit tall for some of us. Some of the detail work left a little to be desired, but Geoff sold all that he could lay his hands on and they are now much sought-after relics of Veloce history.

One problem that spares dealers always used to find was that the plain and spherical parts of the clutch thrust race looked as if they should have a nice ball groove which wears in them, instead of being flat, as Veloce intended them to be. A certain amount of trickery often had to be employed when selling the necessary parts to rebuild the clutch. "Are the grooves in the thrust race in good nick?" you would innocently ask. More often than not the reply would be in the affirmative, when you could tell them that the groove meant wear, and they needed to be thrown away!

Another headache for Velo dealer was the introduction of the "Amal" or "Doherty" alloy handlebar levers, the very shapely alloy ones with an integral adjuster in the lever. The problem was that with only 7/8in between pivot and nipple, it made the clutch lift very marginal. The standard steel lever had an 1 1/8in, which gives the clutch that bit more lift. A certain 'pattern' part manufacturer gave Velo dealers a problem in this period. He made some pattern hollow oil bolts, used for rocker drain pipes and feed and return pipes. Trouble was, he made them in 3/8 26 tpi, not realising that Veloce had used 1/8in BS.P (28 tpi), a small matter of two threads per inch difference.

Around this time a Viper Clubman became available for £50; a good buy. But what to do with it? Answer: the high-speed trials put on by the Triumph Owners at Lydden, the MCC at Silverstone and the Vincent Owners at Cadwell Park. The Viper was given coil valve springs and Joe Dunphy 'looked at' the front brake and all was set. This bike gave me a number of satisfying, if not hair-raising, rides with quite a few first-and second-class awards to its credit. The highlight of this period was my first visit to Cadwell Park, where the Vincent Owners were running a 100-mile race as well as their hour high-speed trials. I had been briefed on the right line for Cadwell by Hugh Evans and Ray Knight, both experienced club racers, so practice went quite well.

The 100-miler was the first event of the day. With a field consisting of Vincent Black Shadows, Thruxton Bonnevilles, and similar right down to vintage machines, there was always someone on the circuit to chase or be chased by. Halfway through, and the rain started; more than once my big gallumping trials boots were used to steady things at the hairpin before the flag went out. A good result: 6th overall, first 500 road-going solo. But better was to follow.

A quick fill-up with juice and we were called out for the first up to 500 cc race. Those 67 laps of practice were put to good use as I thrashed the little Viper for all it could give me in a race-long scrap with Malcolm Elgar on a 500 Egli Vincent. I was regularly taking the Velo to 8,000 rpm out of the hairpin in an effort to stay with the Vincent. In the first race I lost by half a wheel but the second, run in appalling rain and wind, was an even closer affair. It was definitely a case of leaving the brain in the toolbox. A bad start meant that my first lap was a bit hectic. Coming out of the Gooseneck there was a solid wall of fishtails

Uncrating Indian Velos at Geoff Dodkin's shop at East Sheen

(Venoms) in front; it was a case of drift right and down the grass to get at Elgar who was off and away; then a few desperate laps before we reeled him in. The last lap was nip and tuck, but when Malcolm wound the Egli on through Chris Curve, it drifted out. The Viper, with its somewhat overweight jockey (short-portly, my tailor calls me), kept itself on a tighter line and I managed to get through the Gooseneck first, from Mansfield it was flat out, forget the rev counter, just aim for the line! I won by a gnat's smidgen; no wonder Cadwell Park is my favourite track. It is also the circuit where I dumped the bike twice.

I was having trouble with a 250 Suzuki Super Six and tried a touch of demon braking at the Hairpin, which is where I found how good a Velo brake was with 'green' linings! The second dumping occurred while I was going round Charlie's, this time two Triumph Daytonas were the problem, I got the knee on the floor on the previous lap and thought "Take it easy next time, lad" The next lap and I was heeding my advice when suddenly knee, hip, elbow, etc. were decked and we were sliding down the track. No problem until I hit the grass and started rolling. Disco lights, sky-grass-sky-grass-sky-grass, and then I stopped with a strange buzzing noise in the ears. Said buzzing noise was the Thruxton, on its side with the throttle, on a broken clip-on, wound full open with the handlebar tucked under the seat!

A Thruxton, built from all sorts of bits was added to the Viper. The better power-to-weight ratio was good, but it never achieved any wins. I can claim one first, though, seeing my own bike spinning down the road in front of me. I had loaned it to Nick Payton for the last race of the day at Cadwell Vincent Owners; his Venom had croaked. Approaching the Gooseneck, there was something amiss, the field in front scattered in all directions and there, on his knees, was Nick, and in front of him was *my* Venom sliding down towards Mansfield!

I teamed with Nick in a British Formula 500 kilo race, again at Cadwell Park. All went well until, near the end of the first day's run (it was a two day race), we broke a cam follower. The bike was pushed into the parc fermé, and when everyone went off to the pub that night we nipped in, changed the cam followers and were ready for the next days race!

At this time Avon Tyres set up their Road Runner Championship and we took part in 1977, the inaugural year. The Thruxton Veeline was at the time the only catalogued faired machine, so we were the only entrant that year wrapped in fibre-glass. Not that it helped much, with some very handy 400 Honda fours (bored to 500) and an extremely wobbly GT380 Suzuki (bored to 475). I could catch it round the corners, but didn't know which side to pass — it was all over the place! The season was spoilt by a last-lap, last-corner spill while I was chasing Eunice Evans on her rapid 500 Honda four, at Wellesbourne. Her Honda was quicker than the Velo on a silencer, but when fitted with a megaphone for open races it was a match for her machine. I could see she didn't like *that* corner and was all lined up to nip through on the inside when I caught some cement dust and dumped it in a big way. No damage to the bike but my shoulder came out of its socket; result, no racing for six months. Later that year there were three Velos in the Road Runner series; trouble was, each one was that little bit different from the others! We had to make sure that no two were taken through homologation together.

About the time I was getting fit,

Geoff Blanthorn, one time editor of the Velo Club magazine, currently rebuilding Velos in America, riding Bill Snelling's MAC/MDD after the Land's End Trial. Yes, the Dowty forks are fitted with springs!

Mansfield Corner during the Vincent Club's 100-miler. The absence of a 500 Egli Vincent (see story) seems to indicate that the shot was not taken during the 500 cc races!

Geoff Dodkin took in a pair of racing Seymour Velocette Metisses. The finish and general attention to detail took my fancy, so one came my way, the other went out to Australia. With Geoff's contacts with the many firms that made parts for Veloce, we were able to secure original equipment silencers, valves, gears etc., and we used to share these with Ralph, who had the engineering capability at Hawthorne Works, Thame to make other items; mutual assistance saved both firms from duplicating supplies. Ralph's daughter Liz is an absolute wizard at Velo part numbers; the phone used to buzz with "Got any K 119" (timing cover oil unions) "Yes but only if you can let me have a dozen M 2/18" (Thruxton inlet valves).

About this time I made a move across to the motorcycling mecca in the middle of the Irish Sea, so it was natural that the Thruxton and Metisse came with me, and my thoughts naturally turned to riding the Metisse in the Manx. It was handy living at one end of the Island and working at the other; it meant I had a solid six months of going to work via Quarter Bridge to Ramsey, completing the lap after work back over the Mountain to Quarter Bridge. It was not necessarily the racing line I was after on those workday rides, merely the knowledge of the sequence of bends. Later on I went for a few laps with Fred Walton, Ralph Seymour's Manx-based rider. 1978 saw the introduction of the Manx Newcomers' Race (they had held Snaefell Newcomers' races in the 1950s) and I drew No. 1 out of the hat for that race.

Practice was fairly uneventful. My course knowledge helped a great deal, and I found that I could hold my own with some of the newcomers. On the last morning of practice I must have had the devil in me. Coming out of Creg ny Baa I gave it a good handful and let the revs rise to 6,000, then 6,200, then 6,400 ; then it gathered second wind and flew round to 7,000, the motor running like a turbine. Every time I grabbed another gear it roared off again until it was reading 6,600 in top, and Brandish was getting pretty close. This worked out to a theoretical top speed of something like 132 mph! *(Please! Don't try this on a standard motor; this one had parallel roller mains, one-piece pushrods and rocker return springs.)*

I was not looking forward to the bump start on race day. The Metisse was geared three teeth up on the gearbox and one tooth less on the rear wheel, so it took a lot of pushing to get it going fast enough to bump start. When the flag dropped I felt that I was halfway to Bray Hill before I hit the saddle, but it fired first time and I led my starting partner down to Quarter Bridge. Just before Ballacraine I was passed by two Yamahas, repassed them through Doran's and held them off until we hit the Cronk y Voddy straight when they came past again. I was with them until Ballaugh Bridge but their top speed took them out of sight down the Sulby Straight. All was going well until I braked for School House, Ramsey (*never did like that corner, there is a bump on the way in that always threw me out of the saddle*), and suddenly it seemed as if the entire entry was streaming past; I must have been holding them up in a big way! The Mountain climb was no friend of mine; either, the Velo gearbox was like a five speeder with the middle ratio missing. You could rev the nuts off it in second, but the instant you hit third it was off the mega. I tried everything, even trying to put a boot in the mega to see if it

Bray Hill on the first lap of the Senior Manx Grand Prix, 1978. "I know it's the first lap because the engine hasn't started to pump oil all over the rear tyre!"

would help it to pull. All to no avail, so I was forced to sit there and count the hours (or so it seemed) until I could get towards Guthries and into third. There was a heavy mist on the Mountain that day and tootling down to Windy Corner I was suddenly aware that I was catching another rider. I couldn't see him, couldn't hear him, but with the visor up I could smell the oil haze that a 'stroker pumps out when it isn't pulling hard! Suddenly I saw someone riding, like I was, on the white line in the middle of the road. As I passed (sounds good, doesn't it) I glanced at his number — 32, blimey, I thought, he must be quick! He was. Within a millisecond of coming out of the mist he blasted past and disappeared into the distance. Well, he *was* Dave Ashton, winner of the 500 cc Newcomer's class. Over the hump I could claw back a few yards and round to Ramsey again I could hold my own, but up the Mountain, that was that....

I never tried pulling the revs that I did in that last practice session, not after getting a rollicking from the Seymour Equipe about the impending disasters if I tried it again. Pity, I reckon I could have gained another place or two, or possibly blown the thing sky high! Twelfth place after all that, with another ride booked for the Senior Manx.

This time I polled No. 100, so it was quite a wait before we were sent on our way. This was proving to be a heavy-breather of an engine, and when we pitted after three laps the catch-tank and rear wheel were full of oil, all of which had to be cleaned up before the scrutineers would let us resume. The result of the lengthy pit stop was to be flagged off after five laps, but I was still credited as a finisher, 53rd out of 62 finishers.

Whatever happened to Veloce Ltd?

With speculation on what might — could — have happened in happier circumstances

The death struggles of this once proud company as it turned desperately this way and that to escape the stranglehold of debt in a market bedevilled first by Government restrictions and then by competition from the Far East make painful reading.

An apt sub-title would be "An Industrial Tragedy".

There is a simple answer to the question in my heading. It was explained honestly and frankly in the directors' final report to the Receiver.

"The directors attribute the failure of the company to the decision in the late 1950s to produce a scooter. The cost of design and tooling for this product was financed by a loan in 1959 of £75,000 of which approximately £45,000 is still outstanding."

"The company's entry into the scooter market was late and the product did not attract public support. It was therefore left to a diminishing motorcycle trade to support the losses of the scooter venture."

You may be content with that explanation and see the Goodman family as nice, sincere and honest folk, idealists in many ways but not really with the swinging scene, who got it very wrong when they tried to make a scooter. If you are a British bike enthusiast and know the proud history of the marque, you would most certainly say "They should have stuck to motorcycles", and up to a point you would be right. But to accept these simple explanations vindicated by hindsight without study of the situation of the time is less than fair to directors, staff and workers who gave of their best.

Until 1946 the firm had been independent of outside finance but tooling and development for the LE and the purchase of special machinery for its manufacture cost some £100,000 and from then on they operated on an overdraft, which of course was not unusual for a manufacturing firm. But good business prudence would have ensured that the £100,000 was recovered from the sales of the machine — amortised is the word — but although 25,000 LE models were made, the development cost was not and could not be recovered without pricing the machine out of the market. Eventually the £100,000 was written off and one is forced to the conclusion that every LE was really sold at a loss although the continued sales, mostly to the Police, rewarded the elder Goodmans psychologically if not financially.

We might never have known the inside story of the death throes of Velocette — there were no boardroom leaks to an investigative Press in those day (and without a scandal not enough public interest) — had not Joseph W.E. Kelly, a member of the Vintage Motor Cycle Club, chosen to investigate the firm's financial, production and marketing policies over the twenty years prior to liquidation, as his thesis for a degree of Doctor at the University of Bradford. As it was a serious academic work, Joe Kelly was given access to boardroom minutes and documents and had full cooperation from the Goodman family, and his diligent research has provided historians and the Saga with a unique fund of information.

It soon becomes clear that although the Viceroy scooter (it was originally the Viscount but another manufacturer had grabbed that name) was blamed for the failure of the firm, the LE started the downhill slide. The 'Everyman' pressure came from the specialised press and elderly pundits and was supported by the everyday spectacle of millions pedalling to work or waiting in bus queues.

Dealer pressure did push the Goodmans into the scooter project and that was understandable when the loyal Velo man who had struggled in vain to sell the LE (and often landed himself with service trouble) saw a young man, possibly once one of his own staff, rent a shop down the road, get a Lambretta or Vespa agency and sell a dozen or more a week with ease because mass media sold them for him and service was simple and profitable.

Dr. Kelly discovered that Veloce were offered a ready made scooter package from a German firm for £1,800. It was called the Rolleta (German for scooter is Roller). Three of the directors tried a sample and found it needed development in roadholding, braking and weight distribution. As motorcyclists, they would. They should have got scooter owners to try it out and scooter dealers to evaluate it.

As the Rolleta had a bought-in power unit they would

have to make one or buy one, so they dropped the idea and plunged into the Viscount/Viceroy project. Originally it was hoped to power it with an air cooled version of the LE unit but that was too expensive. The end product was a clever design but too late, too big, too expensive, not really a scooter at all, as the public knew them. It would have stood a better chance marketed as a motorcycle ('The Motor Cycle of the Future') with a bigger engine (not a twin that sounded like a single).

Mind you, they nearly sold it as a package to India thanks to the business connections of Sir Alfred Owen who had been put in as Chairman (unpaid) by the bank. The deal fell through because to get the necessary support from the Indian Government, the Indian firm had to export and wanted Velocettes to take 1,000 scooters a year. As they managed to sell 500 up to then and sales had dropped to a trickle, they could not agree. That was typical of the projects which promised to slow the decline and ease the growing cash flow problem but ended in disappointment.

It looked as if there was a future for the LE unit as a power unit for refrigerated ice cream vans. A dairy firm in this country was interested and Velocette developed a stationary engine with electric starter, a governor and a centrifugal clutch. But no orders followed. Unable to wait, the ice cream business like the motorcycle business being seasonal, the firm bought a BSA built engine.

Another firm which specialised in petrol driven generators for a variety of users bought 60 LE stationary engines after a prototype supplied had successfully run a 1,000 hour test followed by a further 500 hours. They supplied them to an ice cream firm in Australia who had trouble with them, refused to pay and counter claimed for £12,000 for loss of business and replacing the engines with others. The British suppliers refused to pay Velocette, and passed on the Australian counter claim. It seemed that in accepting the British firm's order they had accepted that firm's insistence on a 12 month guarantee. Twelve months operation in sub-tropical conditions would be the equivalent to at least 100,000 miles on a motorcycle which was too much to expect. In the event the claim, which grew to £10,000 plus costs and could have sunk Veloce there and then, fizzled out because the British firm went into liquidation.

Hall Green, 1929. The gearbox section

Even the Viceroy engine raised great hopes only to be dashed when a Canadian hovercraft company said they would require 3,000 in 1969 and 600 per month thereafter, at home the Lincolnshire hovercraft company, Hoverair, said they wanted 300 (hopefully with a power increase). Charles Udall was called in to extract more power but the Canadian firm could not raise the money to buy any and Hoverair only bought around 100 and then went into liquidation.

On the motorcycle side, an American plan to market 2,000 machines by mail order came to nothing and the Indian Velo project failed after Floyd Clymer's death.

Earlier plans to diversify into work for the aircraft industry, having done so during the war years, proved disappointing. No longer did Veloce have suitable modern machinery or the skilled labour to carry out such work profitably.

Among the fanciful ideas and good ideas that turned into flops one decision turned out to be a money spinner. They bought the spares operation of Royal Enfield when the Redditch firm closed (apart from the big twin which was hived off), giving £20,000 for some £75,000 worth of spares and tooling and in the first year, 1967, made a profit of £26,000 on a turnover of £53,000, whereas the motorcycle business lost £33,000 on a turnover of £328,000. Kelly says the RE spares operation put off liquidation for three years. By 1964, the demand for singles was falling, and Veloce not able to build and stock enough to take advantage of the short buying season (now like many other manufacturers having to buy components cash on delivery). 75% of the motorcycle business was LEs for the police and it was enlarged to 250 cc. Prototypes were made but there was no production. Bertram then put a cat among the pigeons by suggesting they might import Japanese machines. Yamaha was mentioned as a possibility, their range not (then) conflicting with Velocettes. Nothing more was said about Japanese bikes but there were some discussions with MZ.

I consider Bertram did a wonderful job, always backed on the production side by cousin Peter, first as manager of the racing team on a minuscule budget and reworking old works racers to make new ones, then as Sales Manager struggling in an impossible position of not having enough of the bikes to sell (Venoms and Vipers) and with no hopes of things improving. That his almost crazy (by normal standards) of going for the world 24 hour record with an almost standard sports bike was successful, and still stands in the 500 class, is proof that Fate sometimes rewards a trier. It remains a memorial to a firm that always made a bigger mark in the history of motorcycling than could be expected from its size and pocket.

Many decisions that looked sound at the time, like asking Charles Udall in 1957 to design "a 350 cc twin of the most vibrationless type, with five-speed box and full enclosure" as the next development after the scooter, were not really practical because the scooter used up all the money they could borrow. The bank was pressing them to lower their overdraft in view of Government credit restrictions and the finance company, which came up with the money in 1960, insisted on being represented on the Board. When Veloce, even more desperate for cash, wanted to extend the loan by £12,000 the finance company insisted on a second mortgage on three of the director's houses. Nothing more was heard of the 350 twin but in 1961 the new 250 learner limit resulted in Udall being given instructions to design a unit construction 250 machine with chain drive camshaft, alternator and coil ignition. Udall left next year to go to AMC but the 250 did get to a mock up stage. (It is now

The Indian Velo: would a restyled single have sold?

preserved by David Holder.)

In 1959 Bertram, as Sales Manager, was outspoken about the Viceroy scooter which soaked up money and slowed production of singles he could sell. Unless, he said, they could get 400 scooters to the dealers by March 1960 with 200 in reserve at the factory (a reasonable plan for a launch), and he did not think they could (he knew jolly well they could not), the scooter should be abandoned. The Board did not agree but decided to drop the LE (apart from Police business) and the Valiant. Later they gave Udall instructions to design a 500 cc ohc twin "with a clutch outside the final drive". Note that.

Study of the accounts suggests that in the later years with falling production all models were sold at a loss and they would have been better off making spares, not motorcycles.

The sands finally ran out when no one would lend them any money, not even their shipping agents (Schofield Goodman) who had loaned them £15,000 in 1969 at 15%. E. & H.P. Smith who bought Enfields showed interest in buying Veloce as a going concern but pulled out suddenly after their accountants had reported back and then it was a matter of which creditor would strike the final blow. It may indicate the respect in which the Velocette directors were held that no one wished to convene a meeting of creditors but finally Schofield Goodman, who were Veloce exports agents for the life of the firm, carried out the final task.

From a creditor's point of view it was quite a satisfactory solution. With the sale of the factory, plant and stock even the ordinary unsecured creditors got 84p in the £. The share holders, mainly the Goodman family, got nothing.

A sad end to the saga of a family who, to quote Dr. Kelly, "Were a family dedicated to the manufacture of a product over which they enthused rather than a family determined to take the maximum amount out of the business". They certainly didn't take much out; their rewards as working directors were meagre There was a time when Eugene (then Managing Director) had to ask them to accept some delay in their pay cheques to help the company over a difficult period and I doubt whether they ever took as much as a

What would be the market value of this lineup today? Mk. VIII KTTs ready for despatch from Hall Green

skilled worker on overtime. Mind you, the workers were paid below the average rate but were loyal enough when the firm was racing and morale was high. As the old, faithful ones left, retired or had to be made redundant, younger newcomers developed no such loyalty nor could be expected to with a firm that was obviously dying.

"You know what really broke our hearts, my cousin and I, was the thieving that went on. We were building the last batch of Thruxtons and needed carbs for them. They had to come from Spain as Amals here didn't make that size. We had to pay for them in cash at the docks and over a week or two we scraped together the money and sent Don Harrison down to Southampton with the cash. When he got back we had them fitted to the engines and went home that night feeling pleased. When we got in the next day all the carburettors had been stolen. That was the end as far as we were concerned."

Bertram got a job as buyer for Norton Villiers and actually left before the factory closed. There was nothing left for him to do.

Peter waited to hand over to the liquidator. "It was a great feeling of relief and when I started my little gear cutting business with some machinery I bought at the sale I was happier than I had been for many years," he said, adding "I don't suppose we had done all that badly to keep the firm going... after all, Rolls Royce went bust at the same time. The Government rescued them, of course."

We may all have opinions about what should have happened but it is difficult to put oneself in the position of the board of directors of a family firm that had spanned three generations. There is a case for suggesting that it would have been better if the second generation, Percy and Eugene, had retired after the war and gone home to play with motorbikes, leaving the third generation to fend for themselves, which they might have done well (without the LE handicap). If the firm had stuck to the business they knew, sporting singles, they could have made enough money in the sellers market in the late forties and most of the fifties to have updated, restyled and then redesigned the 500. Everyone else seems to think the solution was to have first resurrected the MOV of fond memory and then gone for the chain ohc unit 250. I disagree and so did Bertram when I talked to him long ago. A 250 costs very nearly as much to make as a 500; the difference in material cost is negligible, it's the labour and overhead costs which matter so, unless you are big enough to make it as a loss leader (Veloce weren't), it has to be made cheap and cheerful and in large numbers. Forget the 250, then, even if it starts brand loyalty in teenagers, which I doubt. More likely it can start brand hostility if the first secondhand one they buy is a disaster.

Bertram was on the right lines when he dreamed up the Venom out of the worthy but unexciting MSS. The pity was that it was stuck with the old frame which was obviously a codged up conversion of the old rigid frame, when Velocettes, not McCandless, were the originators of the swinging fork with twin suspension units that changed the shape (with the dualseat they invented) of motorcycles world wide. The second generation Goodmans did not believe roadster bikes should look too much like racers but the third generation knew pretty well what young people wanted. So the Venom engine could have gone into a Mk. 8 type frame with a Mk. 8 type tank. The Venom could have grown a big-fin alloy head and barrel looking rather like the 500 racers. That not too strong crankcase could have got thicker, the swinging rear fork could have been made wider so it would

Personal heater. The 'billy can' is in fact a one-man coke brazier, the workers favourite way to keep warm at Hall Green

Dark satanic mills! Hall Green in the 1930s

take a bigger-section tyre and allow them to space out the rear sprocket enough to allow them to beef up the clutch. An extra half an inch would make a lot of difference to a Velo clutch, especially if they had thought of putting a hole through the gearbox to take a direct-acting push rod on the gate as Bob Higgs did on his Vulcan.

That 'kipper' silencer, stylish enough in 1931, would have gone in favour of a popular shape of the time, probably a megaphone silencer like the one Bertram developed for the Indian Velo. That's what the Venom Thruxton line could have been. The Viper? Available to order at the same price as the Venom, it cost the same to make. From then on the fantasy is mine, the way I see a basically one model range developing into the 80s and 90s. Drawing heavily on the classic lines of the work's 500s, the Mk. II version engine would look more like the racer because it would have an overhead camshaft, though this time driven by a chain hiding in the big-fin barrel and head casting. It would have four valves, too, no more big valves to tangle, and to set off its purposeful looking lines, it would have those beautiful big bore "Y" type oil drain pipes just like a Mk. 8 (even better looking in braided metallic hose!).

The crankcase would be conventional by now with hefty main bearings in thick walls and a conventional clutch and drive line. It is impossible to estimate the sales lost in the past by the clever but kinky Velo clutch set up, but I am convinced it was the greatest single factor in limiting the market share of the firm to a minor position; it was only conceived as an expedient way out of an engineering problem with the two-stroke range around 1919. That and the unfortunate kickstart ratio which separated Velo men from the majority of potential customers. Twelve volt alternator-fed ignition would help the starting problem and by now provision would have been made in the redesign for electric starting. Racers and café racers who bought one in place of a Gold Star could have had the option of a magneto, in place of the coil contact-breaker. The five-speed box would be a semi-unit design bolted to the crankcase. A new frame of so called featherbed layout would be necessary for the Mk. II, based on the last work's racing frames (a prototype actually exists which was made by Reynolds). Would it sell? Well enough I think to keep a small, tight firm in a modern efficient factory (the old factory long gone and sold as prime building land) working steadily and profitably.

To every technological development there appears to be a reaction, not equal, but considerable. Reaction to the almost universal swing to sophisticated multis in the 80s sparked a new interest in big singles in the classic mould.

The last of the traditional big bangers, a redesigned big Velo, could have been the first to make a come back. Would this Mk. II Venom be raced? Oh yes, but not by the works. They would encourage private owners to race them and would put up prize money in incentives for marque races. Factory development would continue and the results would be made available to owners in go-faster goodies.

An impossible dream? Not impossible while there is still so much magic in the Velocette name.

The latest engineering Goodman

IN HIS analysis of the rise and fall of Veloce Ltd. Dr. Joseph Kelly suggested that it might have been wiser for the third generation, the cousins Bertram and Clifford Eugene (Peter), to have been trained as accountants or marketing men instead of engineers.

"Look what happened to the firms who did have men like that on the Board!" retorted Peter promptly when I told him this. The same reply came from a fourth generation Goodman, Simon, son of Bertram.

Simon, too, was trained as an engineer in tool making and die sinking, but went on to take a post-graduate course in business management at what I call the 'University of Real Life'. Not in this country but in the tougher environment of Canada. Disenchanted with fiscal, political and labour conditions (he had been in trouble with unions once through helping his father at the factory at weekends) in this country, he had gone there to start a new life.

Was his decision to get away from it all down to disappointment at the loss of the family firm in which presumably he would have had a future?

"I would be less than honest if I said it didn't upset me at the time but when I look back on the days when my father used to come home and tell us of the family feuds and boardroom politics. I am glad I had no part of it" he says now. In Canada he found there was a need for 'Saw Doctors', craftsmen who sharpened, tempered and set up the giant ganged circular saws used throughout the logging industry. A university course in the art took five years. Simon taught himself.

'Doctoring' saws led him into the chain saw business, selling and servicing them, and on to the multitude of small power tools that need servicing and sharpening. Starting and building up a business like this from scratch in another country really teaches a man business management skills. He might still be in Canada if his father had not persuaded him to come back to take over from him in the BSA Co. which was one of the small companies that Dennis Poore rescued from the NVT collapse. It made lightweights designed by Bertie from mainly Italian components to export to third world countries. Bikes like Brigands and other names unfamiliar in this country.

He took up a project started by his father to build a "winged wheel" to motorize the millions in India who were still walking or pedalling. His version was a 30 cc alloy unit to replace the front wheel of a push bike, a proper little whizzer that could do 40 mph. He took it to Delhi and demonstrated it but the money to build and market it was not available

Simon Goodman with his Thruxton outside the Geoff Dodkin / Goodman Engineering premises near Evesham

there and the natives are still pedal pushing. The prototype makes a handy runabout round the factory buildings that house Goodman Engineering now that he and his father have pulled out of BSA Co., Ltd. Simon still has a soft spot for Velocette motorcycles and the enthusiasm comes through when he talks about them but his feet are firmly on the ground and making Velocette spares is just one of the firm's activities. First product to put them in the news was the reproduction Norton Featherbed frame, jig-built to precise standards under the direction of Ken Sprayson, who made the originals at Reynolds Tubes.

Most of them take reproduction Manx Norton engines but a version has been made to house a Harley Davidson Evolution 1200 cc Sportster engine to produce Simon's idea of a real super bike. Featherbed handling enables Harley power to be used in safety, light weight giving a performance boost, and superb detailing makes it a delight to behold. "We don't plan to make more than a couple of hundred a year. Look on it as a bike in the Ferrari car class. Something for the man who has everything else" says Simon.

Purchase of the Geoff Dodkin business has given Goodman Engineering a foot hold in the retail spares and machine business, and in the Velocette restoration business, and has widened the firm's base. How does he see the future, Velocette wise? Well, he hopes to make more spares, particularly where he and users are not satisfied that those available are up to Velocette standard. An example is the fishtail silencer which he is going to make to original standards.

Will there ever be a new Venom or Thruxton? Technically he, like Holder, says it is possible but it could not be marketed in Europe or America because of environmental 'pollution' legislation.

Sales of spares may slow as existing bikes are restored so he is thinking of developing modifications and improvements to enable enthusiasts to uprate their machines. An improved clutch for instance, electronic ignition, five speeds, rather in the way his father's development of Venom "go-faster goodies" led to the Thruxton. Development like this could lead to a new, lighter frame (the works experimental frame), a four-valve head, fuel injection (the only way to cope with further pollution legislation) — and who knows, eventually a Mk. 2 Thruxton.

He soon comes back to earth. "The Velocette side of the business has to pay its own way though. The Harley project may produce more business now that we have set up Goodman Engineering America in Oklahoma to market them and Harleys are now interested. The American outlet can help the Velocette business as well".

Meanwhile a computer controlled machine is quietly churning out drilled brake discs for a Japanese manufacturer.

Whatever long term future there is for the Velocette motorcycle it will be safe in the hands of the latest of the engineering Goodmans

Numbers game

STUDY of the engine numbers year by year will give a better picture of the fluctuating fortunes of Velocette than words. Compared with the figures of today's multi-national manufacturers the totals are so insignificant as to be pathetic. But achievement in motorcycles cannot be measured in numbers alone.

The tight-knit family who through two generations made motorcycles because they liked motorcycles and motorcyclists achieved successes in competition out of all proportion to their size and wealth, though were never able to pay for top riders. In technical achievement it was the same story. Enthusiasm and dedication produced innovations like positive-stop footchange, throttle-controlled oiling, swinging-fork suspension, dualseats and vibrationless shaft-drive twins. An enthusiastic owners' club and Velocette enthusiasts world wide ensure that the Goodman family and those who worked with them will not be forgotten.

Four-strokes

Model	Year	From	-	To		Model	Year	From	-	To
Model MOV	1933	1	-	431			1946	6001	-	6265
	1934	432	-	1210			1947	6266	-	7466
	1935	1211	-	1749			1948	7467	-	8304
	1936	1750	-	2486						
	1937	2487	-	3089		Model K (ohc)	1925	7	-	141
	1938	3090	-	3609			1926	142	-	566
	1939	3610	-	4116			1927	567	-	1372
	1940	4117	-	4169			1928	1373	-	2467
	1943	4181	-	4182 only			1929	2468	-	3225
	1946-7	6001	-	6252			1930	3226	-	3546
	1948	6253	-	6500			1931	3547	-	3885
							1932	3886	-	4399
Model MAC	1933	1	-	53			1933	4400	-	4871
	1934	54	-	751			1934	4872	-	5608
	1935	752	-	1644			1935	5609	-	6025
	1936	1645	-	2706						
	1937	2707	-	4034		Model KSS/KTS				
	1938	4025	-	5278		(Mk. II)	1936	7002	-	7746
	1939	5279	-	6557			1937	7747	-	8372
	1940	6558	-	6707			1938	8373	-	9149
WD Models:	1940-1	6721	-	6735			1939	9150	-	9203
MAF & MDD	1943	1101	-	12321			1940	9204	-	9243
	1946	8001	-	10050			1946-7	10001	-	10052
	1947-8	10051	-	11135			1947	10053	-	11300
	1948	11136	-	13759						
	1949	13760	-	14435		Model KTT	1928		-	No. 2 only
	1950	14436	-	15891			1929	3	-	181
							1930	182	-	270
Model MSS	1935	1001	-	1512			1931	271	-	337
	1936	1513	-	2259			1932	338	-	444
	1937	2260	-	2996			1933	445	-	505
	1938	2997	-	3811			1934	506	-	550
	1939	3812	-	4520			1935	551	-	618
	1940	4521	-	4678			1938	700	-	39 (?)
	1943-4	4681	-	4688			1946-53	902	-	2000

Two-strokes

Models A, B2, B3	1923	5001	-	5264	
	1924	5265	-	5947	
Models AC, B	1925	1	-	533	
	1926	534	-	913	
	1927	914	-	1228	
Model D	1936(?)	801	-	1082	
	1922	1083	-	1116	
Model E	1921	1500	-	1512	
	1922	1513	-	2253	
Model EL	1922	2254	-	2257	
	1923	2258	-	2321	
	1924	2322	-	2385	
Model G	1922	3001	-	3069	
	1923	3020(?)	-	3628	
	1924	3629	-	3938	
	1925	3939	-	3970	
Model H	1924	1	-	32	
	1925	33	-	329	
	1926	330	-	518	
	1927	519	-	569	
Models U, U32	1928	1	-	1300	
	1929	1301	-	1912	
Model USS	1929	1	-	262	
Model GTP	1929	1	-	34	
1930 model	1929	1	-	86	
	1930	87	-	1948	
	1931	1949	-	2794	
	1932	2795	-	3763	
	1933	3764	-	4484	
	1934	4485	-	4983	
	1935	4995	-	5474	
	1936	5475	-	6019	
	1937	6020	-	6619	
	1938	6620	-	7102	
	1939	7103	-	7541	
	1940	7542	-	7630	
	1941	7629	&	7631 only	
	1942	7632	-	7639	
	1946	9001	-	9247 exported	

Hall Green, circa 1936

The survivors

IN THIS enlightened or more sentimental age when it is the in thing to collect antiques of all kinds, and collectors scour the earth and beggar themselves to secure anything on wheels which is historic, or uncommon or just old, it seems beyond belief that manufacturers in the golden years of British motorcycling should have discarded the very cornerstones of their tradition. But discard them they did, sold them off disguised sometimes as roadsters, or dismantled them completely; never did they tuck them away in some hallowed corner of the works. It was a forward-looking period in the motorcycle industry and the only thing that was interesting was next year's model. Last year's, no matter what it had achieved, was out of date, obsolete. Backward-looking nostalgia, so common today, had not been invented. The more learned student of history may care to speculate on the cause of this change in outlook. I will stick to the main subject which is: whatever happened to the famous ones?

While the more or less straightforward racers were usually sold to up-and-coming riders... the more "special" racers were, the further from home factories preferred to sell them, to reduce spares complications... real one-offs were considered too much of a long-term liability. Not too way out one-offs were, of course, sometime sold to friends of the "family" who could be expected to keep their mouths shut and not be too tiresome about spares difficulties.

I have long held a view that if you want a machine to be remembered you should give it a nickname, and there is ample proof of this theory in Velocette history. "Clara" and "Jack" have been remembered, and will always be remembered by enthusiasts because they were nicknamed by the ace nicknamer himself, Harold Willis.

There was another Velo which he nicknamed with a name that stuck and this helped to preserve its memory for a good many years and gave it the cachet which protected it from a fate worse than death.

The nickname this time is "Roaring Anna", the Willisism for the model he rode into second place in both the 1927 and 1928 Junior... the one which did not suffer from tappet trouble! Why "Anna"? I know not and it is unlikely that anyone else does for the working of the Willis mind in such matters was devious. Nothing to do with the registration number of OP 7931 issued to Veloce Ltd. on 13 June, 1927 apparently, but Willis must have fallen in love with her for in August the registration was transferred to him and remained so until May, 1929, when it was transferred to Messrs. H. C. Webb & Co., Ltd., the fork makers. Probably for experimental purposes, and it may well be it was used to test the strutted KTT fork. Between times it was used by Webb's service man — a Mr. Ward, I was told by Bob Burgess — on his duty trips in the Isle of Man at race time. At one time, it is said, it was sold to someone at Webbs, but later bought back by the firm. The next part of the story was told by Graham Walker.

Graham Walker recalled that when *Motor Cycling* staged a race meeting at Donington in 1939 and put on a vintage race (they didn't call it vintage, of course... "old stagers" was the description of the machines) it was won by a 1929 New Hudson ridden, believe it or not, by Bill Boddice. Willis was among those present and was, according to Graham, heard to remark that if "Anna" had been around it would have won easily (would, in Willisese, have given the field a lap start and had time for a decoke before the rest finished). Without more ado Willis seems to have got on to Webbs and prized "Anna" out of them with the avowed intention of entering her for the next "vintage" race.

Alas, Willis died barely a month later and it was a long time before *Motor Cycling* could stage another race day. "Anna" was, however adopted by Leslie Udall, Charles Udall's brother, then at Velocettes and later at Triumphs, and used for ride-to-work duties. In the works' bicycle park it was spotted one day by a knowledgeable enthusiast, one J.H. Greenwood, whose name became well known to earnest readers of motorcycle books over the years. While it was in his possession he let Graham Walker have a ride on it. It was war-time and "Anna" was disguised (or disfigured) by a lighting set with a regulation headlamp mask (who remembers them?) and an unlovely pre-war type Feridax dualseat. But despite these encumbrances, and the millstone of a Magdyno it was, Graham estimated, still capable of over 70 mph, and that was on Pool with a silencer. It still handled "like a pukka racer".

Marque loyalty and the interest in historic machines fostered by the Vintage Motor Cycle Club, the Velocette Owners Club and various British bike orientated clubs has resulted in the preservation of many of the significant machines mentioned in this book. Not all those listed have been restored or are yet complete but they will be!.

Locations are omitted for security reasons.

1912 Veloce inlet over exhaust unit construction two speed.

1913 Model A two-stroke 206 cc direct belt drive

1923 Work's TT "Beehive" two-stroke 250 cc.

1927 Experimental spring frame KSS "Spring Heeled Jack".

1927 TT machine ex-Willis "Roaring Anna".

1932 Experimental supercharged racer "Whiffling Clara".

1939 Experimental 500 cc supercharged twin racer, the Roarer.

1939 Prototype 600 cc vertical twin roadster Model O.

Survivors from the works TT machines number:

"Dog kennel" 500 cc models	3
Single ohc 500 cc	3
Post-war dohc 500 cc	2
Post-war 250 or 350 dohc	4

Velocette Specials

The Vulcan
The big V-twin that Velocettes could have made (If they had not spent their money on little flat-twins!)

IF YOU are campaigning a 500 Velo sidecar outfit on the circuits and are fed up with big 'Vinnies' throwing grit in your face as they power by, you might be tempted to buy one and join them. An easy option like that was unthinkable to dedicated Velocette man Bob Higgs, so he chose a much harder one. He would make a Velocette 1,000 cc twin. Being a design engineer, he could not take a short cut of clamping two singles side by side to make a monster vertical twin (like the famous Bob Collier, doyen of special builders, once did with a brace of Norton singles). It had to be a V-twin, that being the only way a 1,000 twin could be fitted into a machine that looked like a proper motorcycle and satisfied his requirement for a low centre of gravity. He was only thinking of the machine from a sidecar point of view, that's why he was so anxious to get the c of g low down, even with the full sized wheels he preferred.

This was to be no effort chalk on the floor with engine and gearbox propped up on wooden blocks while plywood engine plates are cobbled up (and some very effective specials have been built like this). Not by the likes of Bob Higgs! It all had to be worked out on a drawing board before he 'cut metal'. 'Still has drawings and patterns you could buy if you think you could build one.' With no short cuts apart from using as many 500 cc Velo parts as possible, it was a long road that only a professional would have followed. Four long years before he got some pals in for the 'fire up'. You can imagine the joy when it fired up straight away on the kickstart. They celebrated with a bottle of Scotch.

"Making the engine was the easy part", he says, "once I had done the drawings and got the patterns (made by a pattern maker in his spare time) and had the castings cast (his father being a foundry man helped) it was simply a matter of working to blue prints." He had of course at the start of the project moved his car out of the garage and bought a big lathe.

"It's the ancillaries that take the time. Like making proper oil pipes", he said, showing me a set of pipes with built-in banjos all tailored to fit and plated so they look like spares for any production bike. Better finished perhaps. Most special builders would have reached down a length of plastic pipe, a few nipples and a handful of worm drive clips. That's just the oil pipes so you can imagine how long it took to design and make the frame, widely-splayed duplex up front merging into spring frame Veloce at the

Proud owner of a unique motorcycle: Bob Higgs and the Vulcan

back. Plus the front forks, his own design of leading links. The rear fork is Velo with the pivot mountings reinforced.

The engine is, as he says, really very simple. He has designed a completely new crankcase, an inch wider than the single so that the MSS flywheels can take a crankpin one inch wider than the single so that a pair of Venom con rods can run side by side. There's a bronze spacing washer pressed on to the shaft to keep them from rubbing together (if it was allowed to spin it might fret the shaft and weaken it ever so slightly). Big ends are standard, of course. The engine is offset 1in in the frame to keep the chain line standard.

Over the timing gear Higgs resisted the temptation to save cogs by using a common big idler (a not too satisfactory Vincent feature, being noisy at best) and used two sets of standard Velo cogs driven by a pair of crankshaft pinions spliced together. The gears are supported by a steel plate in normal Velo fashion. The Higgs timing gear is much quieter than the Vincent one. Well, it should be with those beautifully cut gears and a more conventional upper rocker gear with more generous bearing surfaces. I am sorry to keep bringing the Vincent into comparison but it is inevitable when considering this machine which has a similar basic concept, 'If you've got a single and want more power double it'. Anyway as a one time Vincent I am on record as suggesting it would have been a better bike for many owners (me for one) had it been built by BSA or AMC I should have put Velocette first! Comparative performance is meaningless for so much depends on the state of tune.

Higgs now uses Venom pistons and heads but retains the MSS cams because for his sidecar work he wants torque in the mid range and that he has got, masses of it. Too much really for the Velo clutch, and 'near the bone' for the gearbox.

Having planned it, drawn it and actually built and then nursed it through its teething stages (remarkably little trouble, actually), he is best placed to suggest why Veloce would not have made a V-twin even if they had thought of it.

"They would have had to completely rethink the cycle parts. The gearbox would have needed beefing (they did this on the work's 500 racers) and they would have had to redesign the clutch at last."

That clutch is still the Achilles heel of the Vulcan. In ten years (and 30,000 miles) development more modification and work have gone into the clutch than any other part. In its present state of development with all-metal friction surfaces, bronze against steel plates and bronze blocks in the chain wheel plus strengthened 'gate' and hydraulic operation it is satisfactory if serviced every 1,000 miles or so.

He explains there is so little lift available on a multi plate Velo clutch. "It was OK when they only had three plates " you cannot afford to lose any of it," I have reinforced mine, and my hydraulic operation cuts out more lost motion."

I love his hydraulics despite my reluctance to use such modern technology when it could be done mechanically so easily. Higgs has found there is room in the gearbox shell to run a pushrod right through from the 'gate' to a point on the outside cover just under the filler plug where he has a tiny slave cylinder. A quick thread or a ball ramp or even a little bell crank would do for capable operation. Half the mystery gone from the Velo clutch operation!

The clutch still slips under maximum torque (2,500 - 3,000) unless it's had new springs recently but the advantage of being all metal is that it can be allowed to slip, when it's a bit like having auto transmission. It slips and then having got rid of the oil slinging from the chain it bites again. Determination to use as much Velo metal as possible meant the wheelbase would be too long for a solo.

If the maker had planned it, a box in shape more like the upright Norton box would have made the MSS wheelbase possible. It would have had a clutch able to stand the power, and understandable to boot. Higgs being a very serious sidecar man (how serious you soon know if you accept a ride in the Steib S 501. He doesn't slow for corners, he accelerates, and the outfit corners like a racing kneeler), he likes the long wheelbase (it's 60in).

When you examine it you can see how the rumours started in the 80s that Velocette had made a secret V-twin. What fooled folk who saw it was that although it was obviously Velo based and gave every appearance in detail work of a factory job, it was anonymous save for the stylised cover plate for the ignition contact breaker, or what ever, on the timing cover which just says VULCAN MADE IN ENGLAND. I like the made in England bit. That set it ahead of other 'specials'. Later the glass fibre tank, which isn't a tank but merely a cover instantly removable for the top end and the big air filter which serves both Concentric carburettors, bore a Velocette transfer above a matching Vulcan one.

Fuel is carried in a steel tank slung from the sidecar chassis rail between bike and chair. Keeps the weight low down, and also in that area is a full sized car alternator driven direct end on from the crankshaft. We are now getting down to personal touches of a serious sidecar man and not what any manufacturer would likely to consider. Like hydraulic operation of rear and sidecar brake and a front brake of an AJS 7R racer (after he had thrown away the corroded magnesium hub and replaced it with more enduring alloy). Braking until this brake turned up was a problem. The original Norton Dominator full width brake was not up to the job (cruising speed is anything up to 80 mph with a maximum close to 100) and he would not adopt disc brakes for aesthetic reasons. Apart from a pronounced squeal above the yelp of the tyre the present brake is superb. The squeal alerts pedestrians better than a horn.

Why Vulcan? Well, it's obvious really. Vulcan, God of fire and metal forging, was a name ready-made for a machine with its alliteration to Velocette. A more media-minded firm than Velocette would have registered this name long ago for a super bike. If they had thought of making a V-twin I am sure the failure of Vincent would have warned them off.

No matter. Bob Higgs has, for a modest outlay, £1,000 he reckons plus the cost of a lathe and no charge for four years, hard labour, got a unique machine which is a joy to ride, be it to Scotland or Germany, a demonstration of his skill as a design engineer and a perfect showpiece for his engineering consultancy, Vulcan Services.

In his spare time he is working on improvements to LE and other Velocettes.

The Covel
Another special the factory should have made: the Hedley Cox siamesed ohc 500 twin

BY its very design the ohc Velocette engine has always attracted special building engineers. They have reduced it, enlarged it and because it is so compact and unusually narrow, the bolder among them have been tempted to double it up to make a twin. Several road-going versions have been built over the years and appear to have given fair results though none seems to have survived.

One built for racing has survived to achieve a certain immortality by being preserved by the Velocette Owners Club and is usually quoted as an example of what Veloce Ltd. could (and perhaps should) have done themselves when it became apparent that motorcyclists in the main preferred twins.

Now I hesitate to mention the Roarer and this home brewed special, the Covel, in the same sentence

Fitting a Mk. II series engine into a spring frame was a popular ploy in the sixties and made a pleasant tourer though the power to weight ratio was even worse than with the overweight Mk. II machine. To overcome this handicap Ivan Rhodes built this special with a Mk. 8 barrel and head to the order of enthusiast John Muglestone

The Covel engine: camshafts driven, in non-Veloce style, by a central chain

but they do have something in common. Neither has been tested in anger. Speculation has therefore ensured them a place in history they might not otherwise have achieved.

The Covel was built in the fifties by a young Veloce race shop mechanic, Hedley Cox, who was a skilled engineer, not merely a spanner man. Moreover he was a brave if rather accident prone rider on his own Mk. 8. He held third place in the 1950 Junior section of the Manx until, of all things, the carburettor needle clip let go and the resultant delay dropped him to 37th spot.

His ambition like that of most ambitious young riders was to ride a 500 and it narked him to see works 500 race bikes gathering dust when he would have liked to put one to good use. He had asked if he could borrow one but had been refused. You can see their point of view. He was a good race shop mechanic, more use to them in the race shop than on a circuit.

This rebuff undoubtedly fired up Cox's determination to build himself a 500 and sustained his efforts over the next twelve months. His 500 was a twin made from cut down Mk. 8 parts (mostly from the Veloce scrap bin) and so cleverly contrived that it slotted into his own Mk. 8 bike, allowing him to contest 350 and 500 events by swapping engines. The more you study this clever yet so simple rework of mainly stock components you find yourself saying "Why didn't more people do it... why didn't Veloce do it when they abandoned the 'Roarer' project?

What made the Covel different from the other siamese ohc Velocettes that had been built, and so elegant in its layout, was that he did away with the traditional shaft and bevel cam shaft design and drove his camshafts by a central chain in the modern manner. Apart from the mechanical advantage of driving the camshafts from the centre, instead of having torsional problems by driving them from one end, the absence of external camshaft drives made the unit neat and symmetrical. In fact externally the outer crankcase halves, the offside one shorn of its bevel drive, the near side one sealed off at the crankshaft boss by a plate, almost match.

The two crankshafts were reduced in stroke to 58 mm from 84 mm which meant standard heads, pistons and barrels (shortened) could be used and the over square dimensions, though a bit revolutionary at the time, subsequently found favour.

The crankshafts were linked by an externally splined coupling, the drive to the clutch being taken from the offside engine, and the camshaft drive was taken from the nearside engine via MAC (early straight cut) timing gears, and the final cam chain drive in an oil tight case between the cylinder barrels. If chain drive for a camshaft offended Velocette purists, at least it was not visible. The magneto, now a twin, sat on the normal platform on the left side of the engine and was driven by chain in the normal way from a rehashed half time pinion. The two camshafts were linked by another splined sleeve and an Oldham coupling. An MAC oil pump with its scavenge side enlarged with half of another lubricated both engines.

The gearbox being in line with the offside engine required an extended gearbox mainshaft to bring the final drive sprocket in alignment with the back wheel but this was supported by an outrigger bearing contrived from a spare gearbox end cover and functioning like the outrigger on a Scott three-speeder.

The only snag I see is that the gearbox stuck out rather a lot on the offside and must have reduced ground clearance though this does not seem to have been a problem at the time.

Cox tried the Covel in practice for the 1951 Manx, said it revved to 8,000 rpm and was "very fast". I've no doubt it was. Teething troubles seem to have been to do with lubrication — inadequate scavenging due to a leaky pump and "foreign matter" in the oil ways. Unfortunately he cast himself away when lying 5th in the Junior race with his Mk. 8 engine refitted and spent the next four months in Nobles Hospital, plus two months convalescing. Doctors told him not to race in 1952 and at the end of the year Velocettes, who were giving up racing anyway, made him redundant. He rode a Vincent twin in the 1953 Clubman's TT but came off at the 33rd Milestone and ended up back in Nobles and was not fit for the Manx. He managed a few rides on the Covel on short circuits without proving anything and then emigrated to America. An approach from the Velocette Owners Club resulted in the engine being rescued and displayed as a showpiece... recently it has been fitted into a late type spring frame so that it can be seen to advantage.

Lloyd Bulmer with a selection of trophies won on his KSS oval racer and sprinter. This is a later version of the machine shown in Lou Branch's advertisement reproduced on page 178

Fame being so ephemeral, it is perhaps fortunate that Fate decreed that this brilliant example of special building should not be tested in real anger, when its shortcomings, if any, might have been revealed, and its reputation damned for all time.

I suppose you could say the same about the Roarer.

The Velocette Owners Club

WITHOUT question it was the 1926 TT win that set Velocettes on the way to fame and devotion from enthusiasts but it was to be a quarter of century before a club was formed catering for owners of this make alone; even then it was for the remarkable LE model and another seven years would pass before a club was founded for single cylinder Velos. Perhaps part of the reason is that most motorcyclists were happier to belong to a club based in their locality and catering for all makes. It could be argued that this was a healthier way of thinking and likely to cause swift advances in design as shortcomings were exposed to critical eyes of other people who had not swallowed a myth. Cynics may think it was because no motorbike was really up to travelling far from home... One make clubs did exist long ago but were not so common as they are today. Perhaps they were an offshoot of the manufacturer's sales department?

The VOC was inaugurated on Saturday, 9th March 1957, as a result of talk on the Velocette stand at the 1956 Motorcycle Show, at a pub called The Red House which used to stand in St. John's Wood, reasonably near Lords cricket ground in London. Several of the founder members are still in the club. There was no chairman but four executive posts: Secretary, Treasurer, Social Secretary and Editor. These posts were filled by three men as the proposed Editor cried off and the Secretary, Peter Hatchett, took it on. The Social Secretary pointed out that petrol rationing would restrict club runs, for this was the year of Suez. The club met, and had a headquarters, at the Laurie Arms in central London but by the end of the first year there were District Centres in large towns in various parts of the country and Freddie Frith had become the first Honorary President. The annual subscription was 15 shillings; prices then asked for second-hand Velos varied from £200 down to £5. Mr. Hatchett was a busy man as he ran a technical queries column in the magazine.

Bob Burgess, who had been Service Manager of Veloce Ltd. for many years, made fascinating contributions to the magazine, as he did for the LE Club too. Other informative articles were provided by ex-employees of the Veloce racing department but these were to cause much distress as they were not paeans of unrelenting praise.

Perhaps the most fundamental changes to the Club came about as a result of the demise of Veloce Ltd. in 1971. The VOC founded a company called Veloce Spares Ltd. and it now operates from its wholly owned premises near Leicester. It provides mail order and monthly weekend shopping for Velo Club members.

The Club is now about 1,800 strong, has centres overseas, runs a race meeting at Cadwell Park each year and supports those who still campaign Velocettes in the IoM. There are also various gatherings of polished old bikes, as is the modern fashion, and most of the activities which are to be found in other old British motorcycle clubs. The club magazine is 'Fishtail' with articles of interest and assistance to all Velocette owners.

Current subscription (1994) is £12 plus £2 initial joining fee. Further details can be obtained from: VOC, The Old Chapel, Cheyney End, Huncote, Leics, LE9 6AD.

James Herbert, Editor, Fishtail

The Velocette Owners Club annual rally, held at Stanford Hall each year

Velocette Owners Club Spares Scheme

VELOCE Spares Ltd. was incorporated in England on 8th September 1977. The Company is registered as a Private Limited Company, which means that as the number of shareholders is restricted to fifty, it enjoys certain privileges under the various Companies Acts. These privileges are principally exemptions from certain requirements demanded by the Registrar of Companies of Public Limited Companies (PLCs) which means the Company suffers less red tape and has greater freedom in its affairs.

The Company is a separate legal entity from the Velocette Owners Club and can enter into contracts on its own behalf. The liability of shareholders for debts incurred by the Company is limited to the amount of share capital for which they have subscribed, currently £6,000, and the Company is authorised to issue shares up to a total value of £100,000 to allow for growth. The nature of a limited company protects the directors, other officers and employees from financial loss in the event of bankruptcy. A creditor of the Company cannot obtain money owing by the Company by demanding it personally from the directors, employees or shareholders.

The Company is wholly owned by the VOC with the exception of the single qualification share held by each of the two directors, which is a legal requirement for a director of a Company. It is the policy of the Company that the VOC will always own at least 51% of the issued share capital of the Company and thereby exercise complete control over the affairs of the Company ie control over finance and the policies of ordering and selling of parts and spares.

The trading policy of the Company is that in general it will not supply parts for the S/A models or LE, but will concentrate on parts suitable for the earlier models. There will be exceptions such as Venom valve springs made in conjunction with a dealer and the range of stainless steel parts which is being continually increased. The company will develop new parts, where appropriate, by applying modern materials and manufacturing methods to supersede the old Veloce specification items. Some examples of this are the roller clutch thrust bearing replacing MAS 57 ball bearing, the pushrod gland nut fitted with an "O" ring to reduce oil leaks and replace dangerous asbestos string and the Omega solid skirt pistons. The Company will also tender for parts made available by manufacturers selling off surplus stocks or dealers terminating their interest in supplying Velocette parts.

The Company offers a mail and fax order service which depends on the time that the other commitments of directors and helpers allow. There is no telephone answering service. The premises are normally open once a month on a Saturday and Sunday from 10am for callers. If we can obtain more manpower we hope to increase the frequency of these "Open Days" as they are usually very popular with members.

The major limiting factor on the growth of the Company at present is manpower. Constructive suggestions and help from members of the VOC, particularly in the time consuming job of dealing with the flow of orders, would be most welcome. At our current open days we usually have sufficient manpower, but again, for more open days more people are required to assist.

It is anticipated that more directors will be appointed, due to the increasing workload, and that the Company will, in time, "go public" thereby enabling all members of the VOC to become shareholders and have a direct interest in the running of the Company. In the meantime, however, ordinary members can influence the Company policy by attending the VOC AGM as any change in the policy proposed and accepted at that meeting must necessarily bring about a change of Company policy by reason of the Club's majority shareholding. Also matters affecting the Company's business will be discussed at Executive Committee meetings of the VOC and your Centre Secretary, who has the right to attend, can pass on the views of his centre. These meetings are also open to ordinary members who may speak at them.

Jim Gould

The LE Velo Club

FORMED from a spontaneous rally of riders meeting at Newlands Corner in Surrey in the early summer of 1950, the LE Velo Club has survived the changing values of over four decades. It now has over 1,000 members worldwide and is a £70,000 limited company committed to the enjoyment, riding and restoration of the post-war lightweight models.

Having in its early years established a strong interest in long-distance touring, and also a close relationship with Veloce Ltd., the Club reached an early peak in 1957 of over 800 members. However its subsequent decline mirrored Veloce's own fortunes, and by the closure of the factory fewer than 350 members remained. By 1973 it was apparent that few dealers were going to remain loyal to the L.E. model, and it also appeared that little information and technical expertise had survived liquidation. It began to look as if the Club would not survive.

However, a dedicated Committee resolved that the Club would lead a post-Velocette revival of interest in the LE and established a spares scheme and a historical archive - the former being stocked with parts bought from dealers who were turning away to other marques. By 1985, when the scheme had outgrown its storage space at the Spares Secretary's house, the Club bought a dilapidated building in Walsall, West Midlands and spent the next two years renovating it for a new role as a spares store and clubhouse. Almost all the work was done on a voluntary basis by Club members.

The Club has now become the prominent source of LE spares and original stocks from dealers has been complemented by purchases from many police forces who used to own fleets of these machines for local patrol work. In recent years a large range of newly manufactured parts have been added to stocks, which have been designed by the Club's highly qualified and experienced spares team.

Donations to the historical archive from many individuals, ex-Velocette dealers and members of the Goodman family have made it a major source of restoration information. It included many Veloce drawings, the original factory sales records and a comprehensive register listing nearly 3,000 surviving machines.

Further details can be obtained by writing to the Membership Secretary, Kevin Parsons, Chapel Mead, Blandford Hill, Winterborne Whitechurch, Blandford, Dorset, DT11 0AB.

Bibliography

Publications

Always in the Picture : Bob Burgess and Jeff Clew (Goose & Son - 1971)
Velocette, Technical Excellence Exemplified : Ivan Rhodes (Osprey Publishing - 1991)
My Velocette Days : Len Moseley (Transport Bookman Publications - 1974)
Velocette: A list of models 1905 - 1971 : Dave Masters (D.J. Masters - 1964)
Velocette 1905 to 1971 - An Illustrated Reference : Dave Masters (Transport Bookman Publications - 1976)
Velocette Development History: Rod Burris (G.T. Foulis - 1982)
Velocette Flat twins : Roy Bacon (Osprey Publications - 1985)

Dealers: UK

The Velocette Motorcycle Co., Meriden Works, Birmingham Road, Allesley, Coventry, CV5 9AZ

M. Arscott, Warren Cottage, Tinkers Lane, Tring, Herts, HF23 6JB
Cheshires, 19-23 Prestbury Road, Cheltenham, Glos, GL52 2PN
Goodman Engineering (Geoff Dodkin), Westward, Buckle Street, Honeybourne, Evesham, Worcs, WR11 5QQ
Drury Precision Sheet Metal, Unit 1, Beechings Way, Alford, Lincs, LN13 9JA
K. & J. Gardner, Wolford Heath, Shipston on Stour, Warks, CV36 5RN
N.M. Payton, 16 Cavendish Road, Colliers Wood, London, SW19 2EY
R.F. Seymour, Hawthorne Works, Park Street, Thame, Oxon
Roy Smith Motors, 116-124 Burlington Road, New Malden, Surrey, KT3 4JB
Alec Swallow, 86 Huddersfield Road, Shelley, Huddersfield HD8 8HE
Alan Walker, 41 Holloway Road, Duffield, Belper, Derbys, DE56 4FE

Dealers: Overseas

British Cycle Supply Co., PO Box 119, Wolfville, Nova Scotia, Canada
British Motorcycle Spares, 9-11 Lloyd Street, Wellington, New Zealand
Domiracer Distributors Inc., 5218 Wooster Road, PO Box 26116, Cincinatti, Ohio, 45226-0116, USA
Dynavector Systems Ltd., 16-15 Iwamoto - CHO, Chiyoda-Ku, Tokyo, Japan
ETS Corbeau sa, 10 Rue Achille Martinet, 75018, Paris, France
Ed Gilkison, PO Box 226, 884-2319, Lake Bay, Washington, 98349, USA
Sven Karlsson, Grindtorp, 585-90, Linköping, Sweden
Motorrad Restauration, Christoph Axtmann, Burbacher Str. 2, 76359 Marxelle 2, Germany
Peters Classic Bike Parts, Rijswijkseweg 363, 2516 HJ, Den Haag, Holland

Clubs:

Velocette Owners Club, c/o VOC, The Old Chapel, Cheyney East, Huncote, Leics, LE9 6AD
The LE Velo Club
Membership Secretary, Chapel Mead, Blandford Hill, Winterborne Whitechurch, Blandford, Dorset, DT11 0AB
Vintage Motor Cycle Club Ltd. HQ and Library Service, Allen House, Wetmore Road, Burton on Trent, Staffs, DE14 1SN.
Tel (0283) 40557 - Fax (0253) 510547

Also available from Amulree Publications:

AURORA to ARIEL

The adventures of a Pioneering Manx Motorcycling Adventurer

In an action-packed career that spanned over two decades, Manxman J. Graham Oates carved a unique niche for himself in the world of motorcycling.

Amongst his many and varied exploits were:

WWI despatch rider 1914 - 1916
Motorcycle manufacturer - The Aurora 1919
ACU Six Days Trial competitor 1920 - 1922
TT rider - Ultra Lightweight TT - 1924
Trans Canada pioneer -
first coast-to-coast trip entirely on Canadian soil 1928
Inaugural World`outboard motorboat record holder - 1930
First person to reach the Hudson Bay
on a rubber tyred vehicle - 1932
Multiple International Six Day Trial medallist 1933 - 1939
One of the founders of despatch rider training in WWII - 1939

£5.95

AURORA to ARIEL is available from Isle of Man booksellers or direct from the publishers:
Amulree Publications
'Amulree"
Glen Road
Laxey
Isle of Man.   IM4 7AJ
(Add £1.00 postage for UK or £1.50 surface postage abroad.)